11 12 13 14 15 16 17 18 **19** 20

THE TECH SET

Ellyssa Kroski, Series Editor

#19

IM and SMS Reference Services for Libraries

Amanda Bielskas and Kathleen M. Dreyer

lita

ALA TechSource
An imprint of the American Library Association
Chicago 2012

Printed in the United States of America

Library of Congress Cataloging-in-Publication Data

Bielskas, Amanda, 1974–

IM and SMS reference services for libraries / Amanda Bielskas, Kathleen M. Dreyer.

p. cm. — (The tech set ; #19)

Includes bibliographical references and index.

ISBN 978-1-55570-782-8 (alk. paper)

1. Electronic reference services (Libraries) 2. Mobile communication systems—Library applications. 3. Instant messaging. 4. Text messages (Telephone systems). I. Dreyer, Kathleen M., 1972– II. Title.

Z711.45B54 2012
025.5'2—dc23

2012007204

⊚ This paper meets the requirements of ANSI/NISO Z39.48-1992 (Permanence of Paper).

Thanks to the cast of the TV show Glee
for providing the soundtrack to our writing sessions
and to our families for their love and support.

CONTENTS

Don't miss this book's companion website!

Turn the page for details.

THE TECH SET® Volumes 11–20 is more than just the book you're holding!

These 10 titles, along with the 10 titles that preceded them, in THE TECH SET® series feature three components:

1. This book
2. Companion web content that provides more details on the topic and keeps you current
3. Author podcasts that will extend your knowledge and give you insight into the author's experience

The companion webpages and podcasts can be found at:

www.alatechsource.org/techset/

On the website, you'll go far beyond the printed pages you're holding and:

- Access author updates that are packed with new advice and recommended resources
- Use the website comments section to interact, ask questions, and share advice with the authors and your LIS peers
- Hear these pros in screencasts, podcasts, and other videos providing great instruction on getting the most out of the latest library technologies

For more information on THE TECH SET® series and the individual titles, visit **www.neal-schuman.com/techset-11-to-20**.

FOREWORD

Today's library patrons are communicating via IM and SMS, and forward-thinking libraries are offering to become their regular contacts. The latest in the evolution of virtual reference, IM and SMS reference services allow patrons to get answers instantly, and they establish the library as an integral part of their information-seeking process. *IM and SMS Reference Services for Libraries* is a start-to-finish passport to implementing an IM/SMS reference service at your library. Authors Amanda Bielskas and Kathleen Dreyer walk readers through everything from choosing the software and developing a staff training program to using QR bar codes to market the new service. This essential guidebook outlines how to integrate this new offering into your existing virtual reference services, how to develop staffing and scheduling models, use widgets, and create a mobile version of your service. Everything from how to use a Google Voice account instead of purchasing a cell phone for the library to implementing an intern program to offer extended hours is outlined.

The ten new TECH SET volumes are designed to be even more cutting-edge than the original ten. After the first ten were published and we received such positive feedback from librarians who were using the books to implement technology in their libraries as well as train their staff, it seemed that there would be a need for another TECH SET. And I wanted this next set of books to be even more forward-looking and tackle today's hottest technologies, trends, and practices to help libraries stay on the forefront of technology innovation. Librarians have ceased sitting on the sidelines and have become technology leaders in their own right. This series was created to offer guidance and inspiration to all those aspiring to be library technology leaders themselves.

I originally envisioned a series of books that would offer accessible, practical information that would teach librarians not only how to use new technologies as individuals but also how to plan and implement particular types of library services using them. And when THE TECH SET won the ALA's Greenwood Publishing Group Award for the Best Book in Library Literature, it seemed that we had achieved our goal of becoming the go-to resource for libraries wanting hands-on technology primers. For these new ten books, I thought it was important to incorporate reader feedback by adding two new chapters to each volume that would better facilitate learning how to put these new technologies into practice in libraries. The new chapter called "Social Mechanics" discusses strategies for gaining buy-in and support from organizational stakeholders, and the additional "Developing Trends" chapter looks ahead to future directions of these technologies. These new chapters round out the books that discuss the entire life cycle of these tech initiatives, including everything from what it takes to plan, strategize, implement, market, and measure the success of these projects.

While each book covers the A–Zs of the technology being discussed, the hands-on "Implementation" chapters, chock-full of detailed project instructions, account for the largest portions of the books. These chapters start off with a basic "recipe" for how to effectively use the technology in a library and then build on that foundation to offer more and more advanced project ideas. Because these books are designed to appeal to readers of all levels of expertise, both the novice and advanced technologist will find something useful in these chapters, as the proposed projects and initiatives run the gamut from the basic how to create a Foursquare campaign for your library to how to build an iPhone application. Similarly, the new Drupal webmaster will benefit from the instructions for how to configure a basic library website, while the advanced web services librarian may be interested in the instructions for powering a dynamic library website in the cloud using Amazon's EC2 service.

After watching Amanda Bielskas and Kathleen Dreyer implement an IM and SMS reference service across the Columbia University Libraries system, I knew that they needed to write down everything that they knew for the rest of us. And that's exactly what they did. These two library leaders share their expansive knowledge and experience in *IM and SMS Reference Services for Libraries,* which guides readers through all aspects of launching a successful synchronous reference service in their library. If you're contemplating an IM or SMS initiative in your library, this is a must-read resource.

Ellyssa Kroski
Manager of Information Systems
New York Law Institute
http://www.ellyssakroski.com/
http://oedb.org/blogs/ilibrarian/
ellyssakroski@yahoo.com

Ellyssa Kroski is the Manager of Information Systems at the New York Law Institute as well as a writer, educator, and international conference speaker. In 2011, she won the ALA's Greenwood Publishing Group Award for the Best Book in Library Literature for THE TECH SET, the ten-book technology series that she created and edited. She's also the author of *Web 2.0 for Librarians and Information Professionals*, a well-reviewed book on web technologies and libraries. She speaks at several conferences a year, mainly about new tech trends, digital strategy, and libraries. She is an adjunct faculty member at Pratt Institute and blogs at *iLibrarian*.

PREFACE

IM and SMS Reference Services for Libraries is based on real-world experience implementing both instant messaging (IM) and text messaging (SMS) reference services in a library. As active users of IM and SMS ourselves, it was a natural connection for us to advocate for these services in our library. In talking with students and faculty, it was clear to us that such a service would be utilized. However, we needed to do the assessment work to back up such theories and decide what exactly would work best for our institution.

You'll benefit from what we learned through this process and be able to apply our success to your library. Library directors and managers and public service librarians will find this book most useful, as it provides practical examples for creating and implementing IM and SMS reference services. All librarians will be interested in our experiences and suggestions for providing the best reference service possible.

ORGANIZATION

IM and SMS Reference Services for Libraries covers everything from how to plan for and assess the needs of your library, get staff buy-in and change the culture at your library, organize and implement a staff training program, and create an internship program to extend the hours of the service. Specifically, Chapter 2 covers the types of software and tools available that provide IM and SMS. We analyze the different services and software, so you get an overview of what is available in one convenient location. Chapter 3 describes in depth the assessment and planning phase that needs to be conducted before implementation can begin. Readers will learn how to assess their user population and survey colleagues about their opinions of IM and SMS. Chapter 4 is an important chapter, as it covers the work needed to garner staff and management buy-in. It also deals with implementing change in your organization. The main focus of Chapter 5 is the implementation of IM and SMS reference services. You'll discover how to choose the right software, decide which staffing and scheduling models are right for your organization, set up your service, create widgets, and create a text messaging service using Google Voice. You'll also learn how to implement the technological aspects of the new service, as well as how to create a training program for librarians and to build an internship program to extend hours. Once all of this hard work is done, you are ready to make the service available to your patrons.

Chapter 6 provides suggestions and ideas for marketing and raising the profile of your library's IM and SMS services, including basic website design tips. As part of providing these services, the librarians who are answering the questions should follow certain guidelines to ensure that the best service is being provided. Chapter 7 covers various best practices in terms of how librarians should answer questions and how staff should interact with each on the service.

Following on that, Chapter 8 covers methods for assessing the use of the service and the answers that staff are providing to patrons. *IM and SMS Reference Services for Libraries* differs from other books on the topic because it includes a detailed planning and assessment section. This is critical as such preparation will help ensure a successful service. In Chapter 8 you'll find information on how to conduct an assessment, as well as questions for surveying users and librarians. Chapter 9 provides a glimpse into the future, offering a preview of tools and resources that could be used for reference services in the near future. The "Recommended Reading" chapter provides the reader with a list of the best print and web resources for further reading.

IM and SMS Reference Services for Libraries provides a comprehensive and step-by-step approach to implementing an IM and SMS reference service. Readers should come away with the knowledge and skill needed to create and implement an IM and SMS reference service at their institution. Most importantly, readers should understand their patrons' needs and how they might use IM and SMS reference to interact with the library and know how best to build the service for their users.

1

INTRODUCTION

- What Is IM/SMS/Text Message Reference?
- How Does It Differ from Virtual Reference?
- Who Uses IM and SMS?
- Why Create IM and SMS Services?
- Implementing an IM/SMS Service

Research shows that patrons use IM and SMS as a regular means of communication with friends, family, and colleagues. In order to meet them at the point of need and interact with them via the methods they prefer, libraries must offer IM and SMS as part of their reference services options.

WHAT IS IM/SMS/TEXT MESSAGE REFERENCE?

Before you can begin to provide these new services, you must first understand the terminology. Instant messaging (IM), or chat reference, is a form of real-time direct text-based communication between two or more people using computers or other devices, along with shared clients or software. The user's message is conveyed over a network, such as the Internet. Short Message Service (SMS) is the text communication service component of phone, web, or mobile communication systems, using standardized communications protocols that allow the exchange of short text messages between fixed line or mobile phone devices. In this book we often use "SMS" and "text messaging" as synonyms and are strictly focused on messages that are derived from cell phones.

HOW DOES IT DIFFER FROM VIRTUAL REFERENCE?

Virtual reference encompasses all reference that takes place in the virtual realm, including e-mail, IM, and text messaging/SMS. In other words, it includes patron reference services that do not occur in person or on the phone. Think of "virtual reference" as an umbrella term and "IM" and "text messaging/SMS" as specific services under the umbrella. In this book we are specifically addressing the IM and SMS forms of virtual reference services. E-mail is an important service to offer patrons and should be included in a robust virtual reference service but will not be discussed in this book.

▶ WHO USES IM AND SMS?

The past few years have seen a tremendous spike in the usage of text messaging/SMS. Text messaging is accessible to most cell phone users; as long as the cell phone user has a plan that enables text messaging/SMS, most cell phones are capable of handling this form of communication. Another reason for the surge in text messaging or SMS is the availability and increased usage of devices such as iPhones, Blackberries, and Androids, which make text messaging easier. In 2010, iPhone sales were estimated to be $37 million and will increase in 2011 to an estimated value of $48 million ("iPhone Sales by Year, 2010–2015," *Market Share Reporter*, 2011, Online Edition; Gale, 2010, reproduced in Business and Company Resource Center).

Some groups, such as teenagers, have adopted text messaging more quickly than other groups. A survey by the Pew Research Center's Internet & American Life Project in 2009 (http://www.pewinternet.org/Reports/2010/Teens-and-Mobile-Phones.aspx) reported that 54 percent of teenagers surveyed text daily. Sending text messages to their peers is their preferred method of communication over calling, e-mailing, and face-to-face communication; when talking to their parents, teenagers choose the phone. In addition to the popularity of text messages for teens, another Pew Research survey in 2011 (http://www.pewinternet.org/Infographics/2011/Generations-and-cell-phones.aspx) found that 72 percent of all cell phone owners surveyed rated text messaging as one of the most frequently used features on their phones.

Instant messaging takes place on the Internet and can be a quick way to communicate with others who are also online. A survey done by Forrester Research found that 33 percent of Americans over 18 use instant messaging at least once a month ("Understanding the Changing Needs of the US Online Consumer, 2010," by Jacqueline Anderson, 2010; http://www.forrester.com/). Clearly instant messaging is a popular mode of communication and affords users a quick way to receive answers to their questions.

▶ WHY CREATE IM AND SMS SERVICES?

Clearly there is widespread adoption of instant messaging and text messaging/SMS among a variety of potential library users. As more and more users of different ages utilize these communication tools, it is imperative that libraries also make use of them. For some groups, such as teenagers, text messaging/SMS is one of the primary ways that they interact with their peers.

Furthermore, libraries need to provide services similar to those used on sites across the Internet, as users will look for these options. For example, while a library provides services different from a retail store, patrons are habituated to IM help on retail sites and this habit will be carried to other sites as well. Users are also familiar with using IM to get quick answers at work or to quickly connect with a friend to ask a question. In order to stay relevant, libraries should meet patrons at their point of need by using the tools that their patrons use on a regular basis, namely, IM and SMS.

▶ IMPLEMENTING AN IM/SMS SERVICE

The book will walk you through the steps of implementing an IM and SMS reference service. It is critical that you think through the implementation so that you build a popular and well-used service. This will take more time and effort than just publicizing an IM handle or text message number to patrons, but it will be worthwhile in the end, as more patrons will use the service and will get the best answers possible.

You can choose to offer either an IM reference service or an SMS reference service, or you can implement both technologies and monitor them through one system at the same time. If you are interested in offering your patrons both an IM and SMS solution, we advocate for choosing one system that integrates both technologies. This will streamline your services and most likely make it easier on the librarians monitoring incoming questions by having to monitor only a single system. There are distinct advantages to each method, and after deliberation you may find, for either financial or patron-driven reasons, that you need or want to offer only one type of service.

In this book, because we advocate for a combined IM and SMS system, we often blur the line between the systems. This is most often the case when talking about a system (such as LibraryH3lp) that integrates both types of services. Much of our text actually addresses both services, and you can apply many of our suggestions to either service if you do decide to go with one over the other. However, it is also important to remember that patrons will utilize IM and SMS services very differently. While library staff may see the back end of the service one way, patrons choose to interact with the library via IM or SMS for different reasons, the main reason being most often a matter of convenience.

2

TYPES OF SOLUTIONS AVAILABLE

- Choose Your Options
- Explore IM and Chat Reference Services
- Explore SMS Reference Services
- Determine Pricing Structures

CHOOSE YOUR OPTIONS

There are a variety of options available for a library that is interested in implementing an IM and/or SMS service. It is also possible to use these services in conjunction with each other. This chapter outlines some of the major software solutions that are currently available for libraries to utilize: some are free, others have minimal fees attached to the service, and others are comprehensive reference management tools that can be costly. Table 2.1 highlights major features of these systems as well as some of their pros and cons.

Unfortunately, this book cannot cover all available options, and it is impossible to predict what the next technological advance will bring or when another solution or software will be developed. Remember, no system is perfect; each comes with its own distinct advantages and disadvantages. This and the other chapters in this book are designed to help you choose among the multitude of options that are available; these decisions will need to be made through discussions of key features and functionalities of the different options with the stakeholders in your library.

Additionally, Table 2.2 shows the pros and cons of each type of option to help you choose between offering an integrated IM/SMS system or just offering an IM or an SMS system separately.

EXPLORE IM AND CHAT REFERENCE SERVICES

IM or chat reference is a form of real-time direct text-based communication between two or more people using computers or other devices,

Table 2.1: Software Pros and Cons

System	Pros	Cons
Altarama (IM and SMS service)	► Highly customizable ► Customer support ► Implements system for you ► Variety of solutions	► Expensive
LibraryH3lp (IM and SMS service)	► Relatively inexpensive ► In constant development ► Works with SMS systems ► Multi-staffing possible ► Unlimited number of accounts and queues ► Web-based or client-based system ► Integrates public IM identities on networks such as AIM, Yahoo!, MSN, ICQ, Meebo Me, and Google Talk	► Annual subscription cost ► Needs stable network connection ► Small team (company run by two people) ► Google does not officially work with LibraryH3lp, so SMS through Google Voice could go down
Meebo (IM service only)	► Free ► Widget-flexible designs ► Integrates web-based instant messengers such as AIM, Yahoo!, MSN	► Annoying ads ► Only one log-on at a time ► No saved transcripts
Text a Librarian (SMS service only)	► Can be used by multilibrary cooperatives ► "Text Speak" translator ► Usage reports ► Searchable archive ► Auto-response available ► Customizable answer templates	► Subscription costs ► Some extra costs may apply ► Only for texting ► Responds to patrons via Mosio website only
Trillian (IM service only)	► Works across multiple platforms ► Mobile versions available; supported on many mobile devices (iPhone, Android, BlackBerry) ► Free	► Client based ► Have to download the client program ► Only one log-on at a time
Web-based instant messengers (AIM, Yahoo!, Google Talk, MSN, etc.) (IM service only)	► Most are free ► Secure/encrypted communication ► Some offer text chat recording	► Ads ► Only one log-on at a time ► Some require registration
QuestionPoint (IM and SMS service)	► Available 24/7 ► Provides training and support ► Customizable ► Administrative monitoring ► Has a knowledge base	► Expensive ► Librarians outside your library answer questions ► Many entry points require log-on info from patrons

Table 2.2: Pros and Cons of Using Integrated and Separate Systems

System	Pros	Cons
Integrated IM/SMS (e.g., LibraryH3lp)	▶ Integrated system for library staff, only have to monitor one system ▶ Offer both services to patrons, seamless to them ▶ Might be easier to get buy-in from library staff ▶ Often allow multiple librarians to staff the service at the same time ▶ Scalable, will work with consortial models	▶ Cost money, sometimes expensive ▶ Take time to implement
IM or Chat Reference (e.g., Meebo)	▶ Many solutions are free ▶ Widgets available with many systems, can place behind a log-in screen or not if you choose	▶ No texting available, single point of access ▶ Many free services have ads ▶ Often allow only a single librarian to be logged on at once
Text Messaging (e.g., stand-alone cell phone)	▶ Mobile friendly ▶ Work on any text message-enabled device	▶ Fees often apply for patrons and libraries ▶ Users are not always in front of a computer or able to access the web ▶ All patrons might not have cell phones ▶ Character-limited response (often 140)

along with shared clients or software. The user's message is conveyed over a network, such as the Internet. The following list is a brief summary of some of the multitude of software options that are available for IM reference. Consult Websites in the "Recommended Reading" chapter for additional sources of information on these products and services:

- **Altarama's RefChatter:** The technology that makes RefChatter possible is a combination of open source software components created by the LibraryH3lp developers. This is then bundled with Altarama's implementation, training, and customer support. It can be used in conjunction with Altarama's SMSreference product. Cost: subscription based.

- **LibraryH3lp:** LibraryH3lp is an integrated IM platform written specifically for libraries. It allows multiple librarians to receive chats through its native chat widget or Meebo Me widgets as well as IMs from patrons on other IM networks such as AIM, Yahoo!, MSN, Google Talk, and ICQ. It is designed to be a mobile-device friendly. It provides two different text message/SMS gateways. Librarians in the same administrative domain can transfer chats or text messages to each other as needed. The system provides an administrative back end for service creation, management, and centralized transcript storage, download, and deletion. Cost: subscription based.
- **Meebo:** This system is a multiple-network, web-based interface. Meebo Me is a widget that would allow people to chat with you directly through your website. It integrates many IM channels into a single solution. Meebo enables libraries to communicate in real time with their users. The system is mobile-device friendly. Cost: free.
- **Pidgin:** Also open source, Pidgin works on multiple networks simultaneously, is client based, and is under constant development. Librarians can log in to accounts on multiple chat networks simultaneously. Pidgin supports many features of these chat networks, such as file transfers, away messages, buddy icons, custom smilies (aka emoticons), and typing notifications (indicate when a user is typing to another user). Numerous plug-ins extend Pidgin's functionality above and beyond the standard features. You need to download the client, and it has no ads. Cost: free.
- **Plugoo:** The Plugoo widget would allow visitors to your library's website to contact you in real time directly through an IM account, such as MSN Messenger, Google Talk, Yahoo! Messenger, AOL AIM, ICQ, and Jabber. If your messenger status is online, visitors can send you an instant message through the Plugoo Buddy in your messenger buddy list. You can also use myPlugoo, a simple link (http://www.myPlugoo.com/) that would allow patrons to IM you from any website, e-mail, or blog. You can chat with up to five patrons at a time, but only one staff member can be logged in at once. The Plugoo widget is customizable. Cost: free.
- **QuestionPoint:** QuestionPoint is a complete reference management system, integrating chat and e-mail reference. The software also allows for the tracking of in-person, phone, and chat reference transactions and follow-up, scripted messages, and it

includes a local and global knowledge base. It features 24/7 live help through membership in the 24/7 Reference Cooperative. Cost: subscription based.

- **Web-based instant messengers:** There are a multitude of IM clients, including Yahoo!, MSN Messenger, Meebo, KoolIM, eBuddy, Radius IM, I Love IM, Snimmer, Mabber, EasyMessenger, IMunitive, Messenger FX, AIM, Google Talk, and QQ. This type of communication is generally conducted by either installing client software on computers or working through the Internet. Most of these services are free; many enable only one log-on at a time. With the free ones, you can also set up more than one account and use them across your library. A more detailed IM services comparison chart is available online (http://www.masternewmedia.org/news/2007/05/20/webbased_instant_messengers_a_miniguide.htm). Cost: most are free.

Other Services

- Fire (http://fire.sourceforge.net/), Adium (http://adium.im/), and Proteus (http://www.proteusx.org/)—open source, Mac, multiple network
- LiveZilla (http://www.livezilla.net/)—Customizable widget with Windows client software, with offline chat requests sent via e-mail
- Miranda (http://www.miranda-im.org/)—Multiple networks, thin client, great plug-ins for scripts using spell checker, works well under low bandwidth, open source
- Trillian (http://www.trillian.im/)—Multiple network program, client based, even works with Blackberry and iPhones

EXPLORE SMS REFERENCE SERVICES

SMS reference is a way for patrons to contact the library via text message. SMS/text messages often have a character limit of about 140 characters per message. Multiple messages can be strung together to complete longer thoughts; shorthand is often adopted in text messages (see Tables 2.3 and 2.4 for examples of "IM speak" or slang and emoticons). There are a variety of options available for text messaging for library reference services, and the following list offers just a sampling of them. These services range from free solutions, to purchasing and using your own cell phone to answer queries, to more robust subscription services. It is important to remember that standard text messaging rates will apply for patrons sending messages from their phones.

▶ Table 2.3: Commonly Used IM/SMS Abbreviations

IM/SMS Speak	Translation	IM/SMS Speak	Translation
ABT	About	MSG	Message
B/C	Because	NBD	No big deal
BRB	Be right back	NIGI	Now I get it
BTW	By the way	NP	No problem
CM	Call me	PLZ	Please
CYE	Check your e-mail	QQ	Quick question
DYFI	Did you find it?	ROFL	Rolling on the floor laughing
F2F	Face to face		
FW	Forward	TA	Thanks a lot
IC	I see	THNX	Thanks
IDK	I don't know	TTLY	Totally
IDTS	I don't think so	TTYL	Talk to you later
IK	I know	TYSO	Thanks so much
IOW	In other words	TYVM	Thank you very much
JIC	Just in case	URW	You are welcome
JK (J/K)	Just kidding	VRY	Very
JW	Just wondering	W/	With
LMK	Let me know	WKD	Weekend
LOL	Laughing out loud	XLNT	Excellent

- **Altarama's SMSreference:** This system provides your library with a unique mobile phone number. Text messages received by that number are automatically delivered to an e-mail address that your library specifies. The librarian monitoring that e-mail address creates responses in e-mail, and responses are automatically

▶ Table 2.4: Commonly Used Emoticons

<table>
<tr><th>Symbol</th><th>Meaning</th><th>Symbol</th><th>Meaning</th></tr>
<tr><td>:) :-)</td><td>Smiley or happy face</td><td>:-/</td><td>Skeptical</td></tr>
<tr><td>:-(:(</td><td>Frown</td><td>:|</td><td>Straight face or grim</td></tr>
<tr><td>;-) ;)</td><td>Wink</td><td>\,,/</td><td>Rock on</td></tr>
<tr><td>>:0 :-0</td><td>Surprise or shock</td><td>\o/</td><td>Woo!</td></tr>
</table>

delivered to the patron's mobile phone by text message. It can be used in conjunction with Altarama's RefChatter product. Cost: subscription based.

- **The AIM hack:** This unique solution allows users to send a text message to the library's AIM handle; the library receives the text message as an IM message and can respond through its AIM account. This solution is free but takes effort to educate your users how to formulate their texts to you. See Chapter 5 (p. 59) for a detailed description of and instructions on how to use the AIM hack. Cost: free.
- **Cell phone reference:** Any library can purchase a text message–enabled cell phone with an appropriate messaging plan and use the cell phone to answer text questions via the phone directly. You can share the cell phone among your staff and share the responsibility of answering the questions. Cost: cell phone plan needed; costs vary.
- **Google SMS:** It is possible to send and receive text messages through the Google Voice application ("app") for free. Google SMS services work on every phone that supports text messaging. You can also send and receive text messages from your library's Gmail address and Google Voice account. See Chapter 5 (pp. 56–58) for more on how to do this. Cost: free.
- **LibraryH3lp:** There are several SMS options available through LibraryH3lp. You can use the Twilio SMS Gateway, the Android SMS Gateway, or the Google Voice Gateway. Each of these options requires a LibraryH3lp subscription; the Twilio and the Android options come with additional costs for the SMS messages you use. In all cases a librarian responds via IM as with other LibraryH3lp questions, and patrons are able to text an actual phone number, not a short code/keyword combination. Cost: LibraryH3lp subscription; some additional fees may apply.
- **Text a Librarian: Mosio—Text Message (SMS) Reference Services:** This system is accessible on over 260 million U.S. cell phones using mobile carrier–approved technologies. A single library or multilibrary cooperatives can use the system. A variety of subscription plans at different costs are available. Cost: subscription based.
- **Twilio:** This system provides U.S. phone numbers and APIs for developers to use in creating their own applications. Libraries with programming resources available could roll out their own text reference service. Cost: subscription based.

- **Upside Wireless:** This is an SMS gateway company offering text messaging solutions to businesses. It is currently being used by UCLA Libraries. It works worldwide with multiple solutions available. There is no software to download. Cost: subscription based.

DETERMINE PRICING STRUCTURES

Many of the solutions mentioned previously in the chapter are free. All they take are staff time and development to implement and manage many of them. Other solutions involve costs. Here are some ways to help minimize expenses:

- **Individual library subscriptions:** Many of these solutions work with an individual subscription for a library. The library can have one account to maintain or manage. Many solutions are often free. For those that are free and only enable a single log-on, you might consider setting up multiple accounts to use in conjunction with each other, especially if you think your service will be in high demand.
- **Consortial deals:** Consortial participation can expand hours of service for little or no extra cost. Working with other libraries can also expand the expertise of your service.
- **Pay as you go:** Many of the SMS options offer chunks of text messages for sale, so you are basically paying only for the messages you send or receive. It might be hard to predict how much use your service will get, and the use may also vary during the academic year, making these kinds of solutions more difficult to predict in terms of budgeting.

▶3

PLANNING

- Assess Your Needs
- Create a Task Force
- Conduct an Environmental Scan
- Conduct IM/SMS Reference Staff Focus Groups
- Conduct an IM/SMS Reference User Online Satisfaction Survey
- Conduct Patron Focus Groups
- Review Usage Statistics
- Analyze Your Data and Set Your Goals

▶ASSESS YOUR NEEDS

Students and faculty use IM and SMS/text messaging as a regular means of communication with friends, family, and colleagues. To meet patrons at the point of need and interact with them via the methods they prefer, libraries must offer IM and SMS as options. To create the service that best fits the needs of your institution and users, it is important to perform a full assessment of the environment, including surveying similar institutions.

If your library already offers an IM and/or SMS reference service and you are interested in expanding it, changing software, or introducing a new service model, the first step is to conduct an assessment of the quality of that service in order to understand how changing service models may affect staff and to gather information from users about their IM use and their use of your IM reference services. If your library does not yet provide an IM service, conducting an assessment of your library and users' needs is an excellent way to establish the usefulness of the service from the beginning.

The following sections cover the different steps you should follow in conducting your needs assessment project.

▶ CREATE A TASK FORCE

A good way to accomplish an assessment is to use a small task force that includes public service staff and a member of your institution's web team if possible. If your organization has an assessment librarian he or she would also be an important member for this team. Before conducting any surveys of users or librarians, the task force should meet to decide what questions you are most interested in answering through the assessment process. It is important to note that during this preliminary phase all questions proposed by team members are valid. A possible set of questions is provided below.

Questions for Assessing the IM/SMS Needs of Users and Staff

- Are users satisfied with the service they receive via IM/SMS (if you already have an existing service)?
- Do your users prefer IM as a mode of communication? If so, why? Do your users prefer SMS as a mode of communication? If so, why?
- What time of day would your users prefer to use IM or text messaging?
- How do your users use IM and/or SMS in their daily lives?
- Who do users expect to be on the receiving end of the IM and/or SMS service?
- What would an ideal IM/SMS service be for your users?

These questions will be the driving force behind your surveys and focus groups. They will also provide an excellent benchmark once the project is completed, as the information that is collected through the surveys and focus groups must answer these questions. Once you satisfactorily answer these questions, then you can go on to the next step of implementing a new service.

▶ CONDUCT AN ENVIRONMENTAL SCAN

Gathering information for an environmental scan will identify emerging issues, trends, situations, or potential pitfalls that may affect the future of your IM/SMS reference services. The library world is small, and various implementations of IM/SMS services are already in use in libraries across the country and around the world. Conducting a comprehensive scan of these services will ensure that you gather the information you need for making the right decisions for your library.

Strategies for a Successful Environmental Scan

- Conduct a literature review focusing on the various approaches to IM/SMS services.
- Conduct phone interviews with colleagues at other institutions that use some of the various services and/or models of implementation in which you are interested.
- To ensure that interviewers all collect the same information, prepare a list of questions for each interviewer.
- Consult the websites of major library organizations such as ACRL or ALA. These organizations frequently conduct environmental scans of the larger issues affecting university, college, and public libraries.

▶ CONDUCT IM/SMS REFERENCE STAFF FOCUS GROUPS

An essential part of an assessment project includes talking to colleagues. If you already have an IM and/or SMS service, holding focus groups with the staff who provide the service will yield valuable information. These focus groups will help you understand how the service worked for the staff providing the service, what they would like to change, and what their experiences have been over the past few years. Most importantly, if you are thinking of making changes such as implementing new software or a new service model, these sessions will be a good way to ascertain how the changes could potentially impact your IM/SMS staff and how they provide the service.

Discussion Points for the Reference Staff Focus Group

- What type of training would you like?
- What sort of service do you think we should offer? Why?
- What hours should we offer the service?
- Who should provide the service?
- How should we staff the service: single librarians, multiple librarians, consortial staffing?
- How should we schedule staff for the service?
- What software should we use (provide examples)?
- How does participating in the IM/SMS service contribute to your job and mission as a librarian?

If you do not currently have an IM or text message service, a focus group with librarians and staff is still an essential part of the process. Such a discussion offers a good opportunity to discuss concerns and allay any fears about change.

Some possible outcomes from these focus groups may include the following:

- Staff should become more familiar with the resources available to them.
- Policy clarification is needed on the scheduling process.
- Refresher training is needed on the current system.
- The statistics database should be revisited and revised.
- Staff from other parts of the organization should be involved in providing IM/SMS reference.

These outcomes should provide actionable steps for implementing changes to your current system to improve and expand it. It should also help you understand weaknesses in your former implementation and training and help you avoid them as you move forward with the next round of changes.

► CONDUCT AN IM/SMS REFERENCE USER ONLINE SATISFACTION SURVEY

If you currently provide an IM and/or SMS service, a survey of your users is an easy and simple way to understand what they think of the service (see the sample survey on pp. 17–18). SurveyMonkey is a great online tool for this. SurveyMonkey allows users to create surveys for free; there is also a paid subscription option. See the website for more information (http://www.surveymonkey.com/). One way to implement this is for each IM/SMS staff member to send each patron a link to the survey at the end of the transaction with a brief explanation that the library is assessing its IM and/or SMS service.

► CONDUCT PATRON FOCUS GROUPS

In addition to surveys, hosting focus groups of users and nonusers is another useful way to gather information related to patrons' expectations for the hours that IM and/or text message services are offered, the subject area expertise of the staff, and the technology used to interact with the library. See Chapter 8 for questions that can be used in focus groups with users. In addition to the verbal questions, it is important to ask each attendee to fill out a brief demographic survey.

IM Reference User Online Satisfaction Survey

1. I am (choose one):

- ❑ Undergrad
- ❑ Graduate Student
- ❑ PhD Candidate
- ❑ Faculty
- ❑ Staff
- ❑ Other (please specify): ____________________

2. Which best describes your field of study (choose one)?

- ❑ History and Humanities
- ❑ Social Sciences
- ❑ Science and Engineering
- ❑ Law
- ❑ Business
- ❑ Medicine
- ❑ Multidisciplinary Studies
- ❑ Other (please specify): ____________________

3. Age (choose one):

- ❑ 18–22
- ❑ 23–29
- ❑ 30-39
- ❑ 40-49
- ❑ 50–59
- ❑ 60–69
- ❑ 70+

4. Is this your first time IMing the library (choose one)?

- ❑ Yes
- ❑ No

5. The resources suggested by the librarian were useful (choose one).

- ❑ Strongly disagree
- ❑ Disagree
- ❑ Agree
- ❑ Strongly agree

6. Overall, using the IM service was easy (choose one).

- ❑ Strongly disagree
- ❑ Disagree
- ❑ Agree
- ❑ Strongly agree

7. I would IM the library again (choose one).

- ❑ Strongly disagree
- ❑ Disagree
- ❑ Agree
- ❑ Strongly agree

8. How did you learn about the library's IM service (choose all that apply)?

- ❑ The library's website
- ❑ A friend
- ❑ A classmate
- ❑ A professor
- ❑ TA
- ❑ A librarian
- ❑ Advertising (poster, flyer, LCD panel, brochure, etc.)
- ❑ Other (please specify): ____________________

9. How do you prefer to contact the library (choose all that apply)?

- ❑ E-mail
- ❑ In person
- ❑ IM
- ❑ Phone

(Continued)

(Continued)

10. Why do you prefer this mode of communication? ______________________

____

11. Where were you when you IM'd the library today (choose one)?

- ❑ In the library
- ❑ At home (or in my dorm)
- ❑ In my office
- ❑ Other (please specify): ______________________

12. What time of day are you most likely to want to IM a librarian (choose one)?

- ❑ Morning (8 a.m. to noon)
- ❑ Afternoon (noon to 4 p.m.)
- ❑ Evening (4 p.m. to 8 p.m.)
- ❑ Late night (8 p.m. to midnight)
- ❑ Overnight (midnight to 8 a.m.)

13. I am satisfied with the hours that the library's IM service is offered (choose one).

- ❑ Strongly disagree
- ❑ Disagree
- ❑ Agree
- ❑ Strongly agree

13. How can we improve the library's IM service? ______________________

____

REVIEW USAGE STATISTICS

If you already have an IM and/or SMS service in place and are interested in changing or expanding the service, analyzing data you have already collected can help. For example, you can use transcripts from IM chats (if you have access) to help assess the types of questions you are currently receiving. If your IM system does not automatically collect transcripts you can often build a system in-house. For example, Meebo (http://www.meebo.com/) is a popular system used by libraries and other organizations to provide IM reference (see Chapter 2 for an overview of available systems). It does allow you to collect transcripts of chats with buddies, but it does not have a reporting system. A simple solution is to have librarians staffing the service fill out a SurveyMonkey form about the transaction and then copy and paste the transaction into the form (see pp. 19–20 for a sample survey).

When we did this analysis at Columbia we discovered that we were receiving a high number of health sciences–related questions but had no health science subject specialists participating in our IM service. One outcome of this analysis was an attempt to get broader subject specialty representation in our IM/SMS workforce.

IM Statistics Survey Using SurveyMonkey

1. Which IM service are you staffing?

____

2. Protocol (choose one):

- ❑ AIM
- ❑ Google Talk
- ❑ Meebo
- ❑ MSN
- ❑ Yahoo!
- ❑ Unknown
- ❑ Other (please specify): ____________________

3. Date and time of transaction:

Month __________________
Day __________________
Year __________________
Time __________________

4. Copy and paste the text of the question here (do NOT include screen names):

5. Question type (choose ALL that apply):

- ❑ Directional: Questions requiring only locational knowledge to answer.
- ❑ E-resources Problems: Questions about connecting to e-resources. Includes technical problems with e-resources.
- ❑ Using E-resources: Questions about using and searching e-resources.
- ❑ Holdings/Known Item: Questions specific to [your library's] holdings. These are inquiries about a known item and questions about why we don't have something.
- ❑ Policy and Procedural: Questions requiring the explanation of a policy.
- ❑ Other (please specify): ______________________________

6. Subject (for "research" and "subject" questions only; choose all that apply):

- ❑ Archives and Special Collections
- ❑ Area Studies
- ❑ Architecture and Fine Arts
- ❑ Business and Economics
- ❑ East Asian
- ❑ Education
- ❑ Health Sciences
- ❑ History and Humanities
- ❑ Science and Engineering
- ❑ Social Science
- ❑ Theology
- ❑ Other (please specify): ____________________

7. Were any follow-up actions required for this transaction?

- ❑ No
- ❑ Referral to subject specialist
- ❑ Referral to Access Services
- ❑ E-resources problem report form
- ❑ E-mail
- ❑ Other (please specify): ____________________

(Continued)

(Continued)

8. Patron status (if disclosed):

- ❑ Undergraduate Student
- ❑ Graduate Student
- ❑ Faculty
- ❑ Non-[Your Institution] Researcher
- ❑ Staff
- ❑ Unknown
- ❑ Other (please specify): ______________________

9. Length of transaction:

- ❑ Less than 10 minutes
- ❑ More than 10 minutes

10. Your comments about the transaction (optional):

11. Patron's comments about the transaction (optional):

There are other ways to gather good usage data in-house in order to gauge the relevance of your service. At our institution we gathered information about IM transactions in a number of ways. Over time our statistical collection mechanisms changed as the software we used changed and as the tools to capture statistics advanced. We instituted an IM service in 2002 with the software system LSSI followed shortly by LivePerson. Both systems captured transcripts. Around 2006 we began using Meebo as our IM service provider; staff were required to manually record IM transactions in a networked MS Access database. After about two years of entering data into that database we developed a SurveyMonkey webform (see pp. 19–20) that tracked more detailed data fields than the MS Access database had previously. Both of these methods relied on IM staff remembering to log each transaction. As this method allows for human error we feel that our usage numbers were regularly underreported. It should be noted that the number of transactions reported went up when we switched to the SurveyMonkey form, as it was easier to find and use, increasing the likelihood that staff would enter the information about their IM transactions. However, we still felt that transactions were underreported.

▶ ANALYZE YOUR DATA AND SET YOUR GOALS

After you have conducted a full assessment, you will need to analyze the results and set your goals. Possible results based on the data collected from the surveys, focus groups, interviews, and analysis of the IM/SMS statistics and transcripts may include the following:

- IM/SMS reference services should continue to be offered (if there is an existing service).
- Service hours should be expanded to reflect the times of day that users conduct academic work, mainly in the evenings.
- Accessing the services should be much easier, and links should be more prominent.
- Points of access need to be increased:
  - Users should be able to IM/SMS the library from many of the high-traffic webpages that they use (or find the appropriate number to text message).
  - Librarians should include the IM/SMS message service in all of their outreach efforts.

Once this assessment is completed and you have a list of actionable results, you should create a list of measurable goals based on these results (see the sample goals provided below). This will be needed when you assess the service once it has been rolled out to the public (see Chapter 7).

Goals for Your IM/SMS Service

1. Provide an IM and/or text message reference service that is both easy for patrons to use and gives high-quality answers.
2. This service receives at least [insert number] questions in the first year with an increase by [X] percent in the second year. If you already have a service, the goal is slightly different: Increase usage by [X] percent in the first year and [X] percent in the second year. [The metrics you use will depend on your organization.]
3. The hours meet the needs of patron, that is, the service is available when patrons need to use it the most.

▸ 4

SOCIAL MECHANICS

- Change the Culture
- Get Management Buy-In
- Maintain the Change

▸ CHANGE THE CULTURE

For some libraries and librarians, instituting an IM and/or SMS reference service may be seen as a significant change in cultural norms. It is important to be aware of the fears that such a change may elicit and work to overcome them. For example, staff may be comfortable with conducting reference in person or via e-mail but may feel that IM or SMS represents a major shift in their work. Often librarians are concerned that they must respond within seconds (quick is better, certainly) and that their answer must be perfect. The biggest challenge will be to understand your colleagues' concerns and allay their fears; see Table 4.1 for some common concerns and ways to combat them. This section describes three steps for overcoming resistance to change:

1. Communicate clearly the need for change.
2. Outline the steps involved to make the change.
3. Find key stakeholders and decision makers to be your champions.

Communicate the Need for Change

Change often fails because staff do not understand the need or the purpose for the change. With careful communication and providing as much information to staff as possible, this should not be an issue. We suggest two strategies for overcoming resistance to change, adapted from a *Harvard Business Review* article, called "Choosing Strategies for Change," in which Kotter and Schlesinger describe several

Table 4.1: Overcoming Resistance to IM/SMS Reference

Common Concerns	Translation	Arguments to Combat These Aversions to Change
Patrons use "IM/text-speak," and I don't know how to respond.	I don't know what they are saying, and I am too embarrassed to ask.	Patrons may use "IM speak," but we don't have to respond in kind. Still, become familiar with the common shortcuts and emoticons that people use, and keep a cheat sheet nearby for reference. (See Tables 2.3 and 2.4 for lists.)
I need to answer questions quickly.	I feel uncomfortable with the pressure. I need time to think!	Patrons are often multitasking and do not expect an answer within seconds. An initial response that you are reading the question usually suffices. If you need more time ask for the person's e-mail and say that you will contact him or her soon with information.
IM/SMS interferes with my daily workflow. I don't have time.	Change is scary.	Change can be difficult, and additional work may seem daunting. However, providing IM/SMS reference will not take up too much of your time because the work will be shared among the library staff members. (This will depend on your staffing model; see Chapter 5.) Remember that IM/SMS reference is an important access point for users who employ these tools on a regular basis. We need to provide this service to stay relevant to our users. (Reiterate your reasons for implementing IM/SMS services to help convince this group.)

methods for implementing change in an organization (see Articles in "Recommended Reading"). The strategies are (1) education and communication and (2) participation and involvement. Table 4.2 describes some of the advantages and disadvantages to each strategy. In our scenario of implementing an IM and/or SMS reference service, it is reasonable to assume that a combination of the two strategies will be sufficient for success. Using both strategies will probably be the most productive in terms of garnering staff buy-in.

Education and communication can be more time intensive than other methods. It involves "getting the word out" throughout your organization about the proposed changes, which can take time and effort. But it is important, as it is an excellent opportunity to correct any misinformation about the new initiative. As with any change, there will be rumors and concerns among the staff, but talking to staff before anything is implemented will prevent false information from circulating too extensively.

▶ **Table 4.2: Change Implementation Strategies**

Method	Application	Advantages	Disadvantages
Education and communication	Used where there is a lack of information or there is inaccurate information or analysis available.	Once the correct information is presented and people are convinced, they will most likely help implement the change.	This method can be more time intensive the more people who are involved.
Participation and involvement	Used when the initiators do not have all the information they need to implement the change; and when the staff who are needed to be involved in the change have considerable power to resist.	Once you get people to participate they will be committed to helping implement the change. They can supply additional information needed to put forth the change.	This method can also be very time-consuming as it can take time to get staff committed to the change. Also, it can slow down the change process if the information the staff provides is incorrect.

Source: Adapted from Kotter, J.P., and L. Schlesinger. 2008. "Choosing Strategies for Change." *Harvard Business Review* 87, no. 7: 130.

Participation and involvement is another useful strategy for implementing this kind of change. In the case of implementing an IM and/or SMS reference service, initiators will most likely have all the information required to implement the service and get it started. The entire staff will not be designing the change, as a small group will most likely lead it. However, because so many people will be needed to staff the service, some of them could contest the change. Therefore, getting staff involved from the beginning is critical. One way to do that is to have the change approved through an existing central committee of colleagues—consisting of colleagues, not management. Having the new service approved by such a group could make staff on the committee feel invested in the change and will encourage them to provide information to others. This type of sign-off may be more important in large organizations than in smaller institutions.

Hold several general meetings at different days and times (see p. 26 for a sample agenda) to introduce the new service and allow staff to ask questions. Demonstrate the new service if possible. Provide documentation during meetings and after via e-mail so that staff have the correct information.

Agenda for Service Change Meetings

- Describe the changes and why they are occurring (see the following section for more information).
- Explain what you need from the staff:
  - Participation in the service (see Chapter 5 for staffing models)
  - Help marketing the service
- Introduce the new service:
  - New software (demonstrate if possible)
  - Hours of the service
- Solicit questions:
  - Leave enough time to take questions from staff.
  - Use this time to correct any misinformation circulating among the staff.

Combining the two methods to use as part of your change strategy requires time and planning. These steps can take longer than just making the changes but in the long run will be more beneficial to the institution. Providing staff with information from the beginning and involving them in the process will encourage them to take ownership of the change and result in a more successful service.

Outline the Steps Involved to Make the Change

As part of the communication and education process, provide staff with the steps and timeline for the implementation as soon as possible. This requires that they be worked out well in advance by the committee or people leading the change so that staff have plenty of advance notice. Staff will need to know that they will receive training if new software is being introduced. Additionally, staff will need to know if and how this change will affect their daily work flows. For example, if staff will be required to alter their schedules to staff the service, this will have to be made clear from the beginning.

Sample Timeline for Implementation

- **Fall 2011:** Complete evaluation and implementation of new software.
- **Spring 2012:** Introduce new service model (change in hours, new software, etc.) to the librarians and staff by holding several general meetings.
- **Late Spring 2012:** Train existing staff on new software.
- **Summer 2012:** Implement new service model and begin training additional public services staff. Continue to hold meetings and training throughout the summer.
- **Fall 2012:** Roll out fully implemented service model.

Find Key Stakeholders and Decision Makers to Be Your Champions

Having management's support will be critical to the success of your initiative (see the next section), but finding peers to advocate for the change may be even more important. The change strategy recommended here requires a great deal of communication and staff involvement so that staff do not try to subvert the new service or the changes being made.

Identifying the people in your organization who will be good advocates for change should be simple; find the staff who will be most impacted by the change. For example, reference librarians and other staff involved in providing the service will need to be heavily invested and will be an important group to get behind the change.

Possible Stakeholders

- **Public service librarians**
  - This includes access and reference librarians and other staff.
  - Champions will be those who have had influence on other changes in the organization or those who are involved in many committees and activities.
- **Managers of public service points**

▶ GET MANAGEMENT BUY-IN

Decision makers will be managers or team leaders who could have a great deal of influence over the success and implementation of the change. It will be critical to get management on board before introducing the changes to the stakeholders and other staff. Before approaching management, filling out the change planning form (see the sample form on p. 28) is a helpful step. This form will help you articulate the change, the reasons for it, the goals and outcomes, and any possible resistance to the change.

▶ MAINTAIN THE CHANGE

Creating and implementing a new IM and/or SMS reference service is not the end of the change process. Once the service is implemented, the biggest challenge will be to maintain and grow the service. Assessing the service will be a useful way to decide how and if to improve what is offered and to gather feedback. Sustaining the change will depend on if you were able to change the culture and

Change Planning Form

(A blank form is available on the companion website at www.alatechsource .org/techset/.)

1. What is the change?
   Implement an IM and/or SMS reference service at the library.
2. Reasons for making the change?
   IM and SMS reference are important tools used by people from all walks of life to communicate with others. Allowing patrons to contact a librarian via IM and/or SMS is becoming a mandatory service. Research shows that Generation Y uses IM and text messaging more often than e-mail.
3. Desired outcome of the change; that is, what are the goals that you would like to accomplish with this change *(goals from Chapter 3 included here)*?
   a. *Provide an IM and/or SMS reference service that is both easy for patrons to use and gives high-quality answers.*
   b. *The IM and/or SMS reference service receives at least [insert number] questions in the first year with an increase by [X] percent in the second year. If you already have a service the goal is slightly different: Increase usage by [X] percent in the first year and [X] percent in the second year. [The metrics you use will depend on your organization.]*
   c. *The hours meet the needs of patron, that is, the service is available when patrons need to use it the most.*
4. Who will champion the change?
   Those most affected by the change should make good champions; they include public service librarians and other staff involved in public service work.
5. What forces may be working for or against the change?
   ***Forces working for change:** compelling reasons to change; advocates of the change*
   ***Forces working against the change:** resistors to change [see Table 4.1 for possible negative reactions to implementing an IM/SMS reference service and how to combat them]*

Source: Adapted from Warrick, D.D. 2009. "Developing Organizational Change Champions." *OD Practitioner* 14, no. 1: 14–19.

create an environment that is more receptive to providing IM and/or SMS reference. Constant communication with staff about why it is an important service to offer, even after implementation, will be a critical component to your success.

In "Developing Organizational Change Champions," Warrick (2009, *OD Practitioner* 14, no. 1) identifies three stages of change that can be applied to the process of setting up and maintaining an IM service, which we address throughout the book:

- Stage 1: Preparing for change:
  - Create goals (see Chapter 3).
  - Get staff and management support (covered earlier in this chapter).
  - Educate the staff about the change (covered earlier in this chapter).
  - Train the staff on the software that will be used (Chapter 5).
- Stage 2: Implementing the change:
  - Evaluate software (Chapter 2).
  - Choose hours and a staffing model (Chapter 5).
  - Set up the software (Chapter 5).
- Stage 3: Sustaining the change:
  - Assess the service, including gathering feedback from staff (Chapter 8).

5

IMPLEMENTATION

- Evaluate and Choose an Application
- Develop Staffing and Scheduling Models
- Develop a Training Program
- Create an Internship Program
- Create and Use Widgets
- Use Google Voice for Implementing an SMS Reference Service
- Utilize AIM SMS Hack
- Resolve Spam Issues
- Create a Mobile Version of Your Service
- Create an IM Library Kiosk

Once you have completed your assessment and received support from your colleagues and management, the next step is to finalize the right software for you. Once you have made this choice, the implementation process can begin. If you do not have an IM/SMS reference service already in place, this process may seem daunting at first. However, after having gone through the assessment and preparation work outlined in previous chapters, you will be well prepared to make decisions on how to proceed with the next phase. Alternatively, if you have a program in place but are just changing the software and/or changing the staffing model, some of the steps outlined in this chapter may seem familiar.

No matter which piece you are implementing, as emphasized in Chapter 4, communication will be essential. Staff will need clear outlines of what is being implemented and what their role is in the process. The more information you can provide staff as you move through the various stages, the more invested staff will feel and the better the service will be.

In addition to implementing the software you use, there are other components of the program that will need attention, including developing an internship program and training your staff to use the new system. Often you will need to work on several of the projects discussed in this chapter simultaneously. The projects outlined in this chapter will walk you through various steps of implementing your IM/SMS service. However, keep in mind that each system is different, and you might not be able to use all of the following projects with the system you choose. Additionally, it may also be possible to add some of the following technologies to supplement the service(s) you already offer.

▶ EVALUATE AND CHOOSE AN APPLICATION

The first step to implementing an IM/SMS reference service is to choose the best software for your organization. Chapter 2 compares many of the main IM and SMS systems currently available. In short, you will need to create a list of features that the software needs to have; this will come out of your assessment of users and staff (see Chapter 3 for more information on how to do this). Then you should evaluate different software, keeping these requirements in mind. Once you have chosen the right tool, you will need to implement the software.

Deciding which IM/SMS system is right for your organization depends on many factors. To help choose between the different systems, a description of the major ones and a list of pros and cons are available elsewhere in this book (see Tables 2.1 and 2.2, pp. 6 and 7, in Chapter 2). Additionally, during the assessment of your library's needs, when talking to or interviewing library staff, it would be worthwhile to survey library staff and ask them about their desired requirements for an IM/SMS system.

The checklist on pages 33–34 contains potential IM system requirements that you can use to help determine the desired features of a system. Determine what you absolutely need the system to be able to do (must haves) versus what you would like it to do if possible (should haves and nice to haves). You can balance this checklist with the properties and pricing of each system to help you choose the solution or software that will work best for you.

If there are currently no IM/SMS systems that match your requirements exactly, you might have to compromise on your wish list of features or go with the next best solution that is currently available.

IM/SMS System Requirements

Must have	Should have	Nice to have	
			Functionality
❑	❑	❑	▶ More than one librarian can access at a time
❑	❑	❑	▶ Reporting functions (statistics; must be exportable to .csv or tabular format)
❑	❑	❑	▶ Visible queues, status, and assignment for all staff looking at the data
❑	❑	❑	▶ Must be reliable
❑	❑	❑	▶ Queues and the ability to move between them
❑	❑	❑	▶ Spell check
❑	❑	❑	▶ Works on mobile devices
❑	❑	❑	▶ Can send images and files
❑	❑	❑	▶ Can send messages to cell phones and other devices
❑	❑	❑	▶ Works with text messaging systems (or includes one)
❑	❑	❑	▶ Capable of easily transferring a query from one librarian to another
❑	❑	❑	▶ Could have multiple instances (e.g., different queues)
❑	❑	❑	▶ Can handle IMs from commercial providers (e.g., Google Talk, AOL, etc.)
❑	❑	❑	▶ Librarian-to-librarian interactions possible
❑	❑	❑	▶ Functional from reference desks
❑	❑	❑	▶ Does not require librarian to be on the local network (available from anywhere)
❑	❑	❑	▶ Has canned answers available automatically
			Privacy
❑	❑	❑	▶ Ability to store data disassociating patron names/IDs from content of questions
❑	❑	❑	▶ Archives and data must be stored on local server(s) or secure vendor server

(Continued)

(Continued)

Must have	Should have	Nice to have	
			Privacy *(Continued)*
❑	❑	❑	▶ Supports patron authentication
❑	❑	❑	▶ Individual identifier for each transaction
❑	❑	❑	▶ Transaction logs must be searchable (keywords)
			User Interface
❑	❑	❑	▶ Customizable user interface
❑	❑	❑	▶ E-mail transcript to user option
❑	❑	❑	▶ Can send clickable links
❑	❑	❑	▶ Cobrowsing
❑	❑	❑	▶ Works on common browsers and operating systems (for users)
			Other Options
❑	❑	❑	▶ Scalability (can grow with your needs)
❑	❑	❑	▶ Support from vendor
❑	❑	❑	▶ Knowledge base for staff and/or public (FAQ builder)
❑	❑	❑	▶ Tagging (categorization) by staff
❑	❑	❑	▶ If it's a commercial product, has good customer service reputation and support
❑	❑	❑	▶ Good documentation
❑	❑	❑	▶ Customizable staff interface
❑	❑	❑	▶ Surveys possible
❑	❑	❑	▶ ______________________
❑	❑	❑	▶ ______________________

Cost will also be a major determining factor; it's important to have an established idea of your budget for the project to help ensure you don't get too excited over potential solutions your library will never be able to afford. The good news is that there are solutions for just about every type of budget (some options are free!).

If there are no solutions you are thrilled with, choose the one that will work best for you at the time. Remember, this doesn't have to be a permanent solution! You can always migrate to other software at a later date. New solutions become available all the time; you never know what the next technological advance will bring.

Chapter 2 provides a detailed analysis of some of the different IM systems available. The assessment of your organization, staff, and users will provide guidance on choosing the right software. After conducting your assessment you will be able to create a requirements document or a list of software and system requirements that is based on the needs of your users and organization. For example, based on your assessment you may determine that you have the following requirements for a new system:

- Allow multiple librarians to be logged in at one time
- Allow chats to be transferred between librarians
- Capture transcripts
- Provide robust usage reports

The requirements document would list these items as priorities, and you should refer back to this document every time you review potential software. The following is a short requirements document; the full list of potential requirements is included in the IM/SMS System Requirements checklist (pp. 33–34). See the companion website http://www.alatechsource.org/techset/) for the blank requirements document to download.

Sample Requirements Document

Must haves:

1. Keeps transcripts
2. Allows multiple librarians to be logged in at once
3. Provides robust reporting
4. Transfers questions between IM staff

Nice to have:

1. Includes knowledge base
2. Allows for multiple instances of the IM widget

Price range: under $2,000 for unlimited number of users

Keep in mind the following:

- You can use the available IM system until a better option presents itself.

- Revisit the programs to see how staff configurations and scheduling, hours of service, and patron use of IM/SMS are working out.
- Continue studying trends and technology, including IM, texting, and so forth—you never know when a better solution will arise.
- Assess the service on an ongoing basis to recommend changes and enhancements or improvements.
- Encourage staff experimentation with new technologies and methods.
- Price is an important factor that can influence all else. It is essential that you understand how much your organization can spend before you evaluate any software that is subscription based.

DEVELOP STAFFING AND SCHEDULING MODELS

As stated previously, in addition to the technical aspects of choosing and implementing new software, the human factor about how it will be staffed must be given a great deal of thought. Your assessment results will help you form an understanding of what your organization and colleagues hope to achieve with the IM/SMS service and how staffing will work. Once you choose a staffing model, the biggest issues that will need to be resolved during implementation will be how you will schedule staff.

Step 1: Choose the Hours for Your Service

Steps to Choosing Hours

Step 1: Review assessment to determine which hours would work best for users.
Step 2: Determine if evening hours will work and if creating a flex-scheduling model for librarians or hiring interns is possible.
Step 3: Be flexible! Assess hours after a few months to determine if they are meeting users' needs.

First you will need to establish the hours for your IM/SMS service. Again, your assessment of users will provide the most pertinent information about what hours you should provide the service. Try to tailor your hours to users' needs and not to the needs that best suit staff. This may be difficult, as students in particular like to work late at night; offering evening hours will be important. You may be able to extend your hours depending on which staffing model and scheduling method you choose (the consortial model lends itself well to this) and/or if you are able to employ interns, adjunct librarians, or provide a flex-scheduling model (covered later in this chapter). It will also be important to try to offer consistent hours so that users know when they can get help using this service. However, some hours are better than no hours, so try to be as flexible as possible when deciding. For example, you can start with a small set of hours and then assess how well they are being used.

Step 2: Choose the Right Staffing Model for Your Organization

Staffing Model Choices

- **Single librarian staffing:** One librarian fields and answers incoming IM questions.
- **Multiple librarian staffing:** Two or more librarians field and answer incoming IM questions.
- **Consortial librarian staffing:** Staffing is shared by several branches, schools, colleges, or universities.

Single librarian and multiple librarian staffing are the simplest to implement, as they involve staff only from your institution. Consortial staffing is more complicated, as it will require coordination with other schools, branches, colleges, or universities. As such, scheduling staff will necessarily be required to minimize any confusion. Careful planning and care must be given to developing policies and procedures across institutions so that users receive consistent and reliable service. Additionally, some IM/SMS software allow for only one type of staffing, that is, only one staff member can be logged in at a time. It is important to keep such factors in mind when choosing your software and considering your staffing model.

After conducting your assessment and choosing your software, you will then need to decide how to implement your service. An important part of this will be how you staff your service. You will have created hours for the service based on your users. In conjunction with this your focus group with the staff will have given you some indication of how they want to proceed with staffing the service, such as multiple librarians or single librarian staffing. Table 5.1 provides the pros and cons for the different staffing options. After weighing the needs of your staff and users, you will be able to choose the right one for you.

▶Table 5.1: Pros and Cons of Different Staffing Models

Staffing	Pros	Cons
Single librarian	▶ Do not need as many staff to provide the service ▶ Easy to schedule staff	▶ Can be overwhelmed during busy times
Multiple librarians	▶ Can provide quicker and more efficient service ▶ Can handle multiple questions during busy times ▶ Are not overwhelmed and can help each other	▶ Need more staff to provide the service ▶ Can be confusing if clear protocols are not in place ▶ Scheduling may be more complicated
Consortial staffing	▶ Extend service hours without additional staff from your institution	▶ Patrons at your institution may not get the best service because staff at other libraries may not be informed about all policies and procedures ▶ The most difficult to implement as it requires coordination between multiple locations

Step 3: Choose a Scheduling Model

Scheduling Model Choices

- **No schedule for staff:** Do not make a schedule for staff; staff log in as they have time.
- **Schedule staff:** Maintain an hourly schedule for staff.
- **Flexible scheduling:** Maintain a schedule, but encourage staff to log in additional hours to extend service times. (See Create an Internship Program, pp. 44–48, for options to expand hours past the usual working day.)

If your organization provides IM/SMS reference from 11 a.m. to 5 p.m., Monday through Friday, for example, it must be decided how those hours will be divided up and covered by staff. A staff member can be responsible for scheduling staff by the hour or for blocks of time, or you can choose to not schedule staff at all. If you employ the no schedule method, clear guidelines and expectations must be created for the IM service. For example, staff will be expected to log in when they are staffing a physical reference desk, thereby providing some consistent level of coverage.

Staffing Your IM and/or SMS Reference Service at a Physical Reference Desk

This is an important consideration and depends on how busy your desk is and how comfortable staff feel managing both services. If both service points are very busy, it may not be possible to staff your IM or text messaging service at your reference desk. Both in-person and virtual patrons can demand a great deal of attention; this decision should be considered carefully with the outcome implemented as a best practice.

Pros:

- More staff may be able to participate in the service if they are not restricted to off-desk hours only.
- Allowing staff at less busy physical reference desks to staff IM/SMS service may help them feel as if they are better using their time.

Cons:

- Staff may be too busy at one service point to respond to users at another service; that is, if the physical desk is busy staff may be too busy to respond to an incoming IM or text question.
- The potential to get interrupted might be high.

If you choose to schedule staff for discrete shifts or blocks of hours, equity among staff will have to be considered. Individuals, divisions, units, and so forth, could be given a certain number of hours a week they must cover so that all areas of the organization share responsibility equally. This is important to decide for scheduling purposes, as staff will need to know how many hours they will be responsible for. For example, you may decide that each staff member is responsible for only one hour a week, but the division, library, or unit may be responsible for ten hours per week. Or, because of staff shortages, each staff member may be asked to contribute multiple hours a week to the service in order to ensure the hours are covered.

If scheduling of staff is the preferred approach, a method needs to be put in place so that staff can indicate which hour(s) work best for them. This can be as simple as putting a call out via e-mail to all participating staff to submit their request. Alternatively, a calendaring tool such as Google Calendar where staff can enter their time slots themselves can be employed to help simplify the process.

Once these issues are worked out it should be fairly easy to pair a staffing model with a scheduling method. The flexible scheduling method may be the trickiest to implement. In this method staff are encouraged to log in outside of their scheduled hours to extend service times. In order for this to be successful, clear guidelines will

► Table 5.2: Pros and Cons of Different Scheduling Models

Scheduling	Pros	Cons
No schedule for staff	► Allow staff to make own schedule ► Do not have to assign a staff member to manage scheduling	► No control over when staff will be monitoring the service ► May not be able to provide consistent service hours
Schedule staff	► Know who and when is staffing the service ► Can provide consistent service hours	► Very time-consuming to manage the scheduling process ► Staff must make sure to find substitutions when they cannot fill their scheduled time slot
Flexible scheduling	► Extend service time without the extra work involved in scheduling	► May not be able to provide consistent service hours; will need to provide clear guidelines on what is expected from staff

need to be provided for staff so that they understand exactly what is expected from them. Perhaps each staff member could commit to logging into the service an additional hour a week before or after the posted schedule times.

See Table 5.2 (above) for pros and cons of the different scheduling models.

Step 4: Pair the Right Staffing Model with the Right Scheduling Method

In order to maximize efficiency for your staff and for users, pairing the right scheduling method with your chosen staffing model is an important decision. Your choice regarding which scheduling method you employ will be informed by your assessment and the staff's feedback regarding this (see Chapter 3 for a list of suggested questions to ask during the staff focus groups).

As discussed in Chapter 4, introduce staff to the new software and IM/SMS reference service in general meetings first so that all staff can ask questions. This will inspire ownership and create buy-in among staff members. It is also a good opportunity to find out about any lingering concerns or issues among staff before the service goes live (see Chapter 4 for more about managing change in your organization).

How to Implement Scheduling Staff with a Single Librarian Staffing Model

Step 1: Inform staff that they will be responsible for all questions that come in during their shift and that you will be scheduling them for discrete hours or shifts.

Step 2: Inform staff of the hours that need to be covered and how many hours they are expected to cover (see earlier discussion).

Step 3: Set up a shared calendaring system like Google Calendar.

Step 4: Ask staff to select their hour(s) on the calendar, or they can e-mail a designated person if that works better. It will be easier on you to have staff select their hours on a group calendar. Hour selection will be on a first-come basis. Give staff a time frame for selection—no more than two weeks—and send out a reminder halfway through the selection period.

Step 5: Once the selection period is over, determine which slots are still available and try to fill those. A call can be sent out via e-mail to all participants asking for additional volunteers.

Step 6: Once the schedule is set, policies should be created for staff about their responsibilities in terms of maintaining their IM/SMS schedule. See Chapter 7 for more about establishing policies.

▶ DEVELOP A TRAINING PROGRAM

Overview of a Training Program

- Before training begins, develop identifiable goals and learning outcomes that you want all those in attendance to achieve by the end of the session.
- The introduction should cover any new policies or procedures related to the IM service.
- Training should be hands-on and focused on learning the new software.
- During training and via e-mail after each session, share the link to an internal website where documentation, canned messages, calendar, and the URL to log in to the service can be found.
- It is helpful to have two people run the training so that one person is available to provide one-on-one help.
- To ensure that as many people as possible can attend, provide multiple sessions as well as individual sessions for those who miss training.

It is critical that all staff involved in providing the service are well versed in the chosen software. Training should be hands-on if possible. Creating a training program with documentation that staff can refer to later will help increase the success of your program. Documentation should include information about how to use the system as well as information about policies and procedures for handling IM/SMS questions. Typically, these policies and procedures will be institution specific and depend on those already in place for handling other reference questions (see Chapter 7 for additional best practices for staff).

Two policies specific to providing IM/SMS reference that may be common to all are:

- View IM/SMS as just another point of contact: an IM or SMS transaction is similar to other public services transactions.
- Employ the same library policies you use when helping patrons in person, on the phone, or via e-mail.

The training should be structured so that it has easily identifiable goals and learning outcomes, such as the following:

- Staff should be able to log in to the service.
- They should be able to successfully answer an IM and/or SMS question.
- Staff should be able to utilize canned messages in their responses when appropriate.
- Staff should be able to use some of the other basic features the software provides, such as transferring patrons to another librarian or e-mailing transcripts to patrons.
- Staff should understand and be able to apply the policies and procedures discussed at the beginning of the session.

Canned Messages

You will quickly realize that many of the questions you receive through your IM/SMS reference service will be repetitive. Some systems may allow you to create "canned" messages within the database so that staff can use those in their response. If that is not possible, creating a list of questions with ready-made answers will be helpful to staff and ensure that users receive consistent answers. These questions and answers can be posted on an internal website where other documentation about the service and software will be kept. See Chapter 7 for more on canned messages.

As you prepare the training think carefully about what you need the trainees to learn by the end of the session and why they need this information. If possible, place all documentation and training information (see the Documentation Checklist, p. 43) on your institution's intranet so that staff can find it easily after the session is over.

Documentation Checklist

- ❑ Basic "How-To" document for signing into software and using other features (two to three pages; see companion website at www.alatechsource.org/techset/ for an example)
- ❑ Best practices document regarding policies and procedures (no more than five pages; see companion website for example)
- ❑ Canned messages for frequently used responses (see Chapter 7 for examples)
- ❑ User manual for software (if applicable or available)
- ❑ Documentation on how to update scheduling system (if using one)

Implement the Training Program

Organization is the key to rolling out a new service and making sure that all who need training receive it. Create a timeline for the training program, starting several months before the training will begin; the following provides an example timeline:

- Three months before service is scheduled to start:
  - Develop documentation.
  - Develop learning outcomes.
- Two months before:
  - Schedule training sessions. (*Best practice:* Schedule multiple sessions on different days at different times of day to ensure as many people can attend as possible.)
  - Advertise sessions to staff.
  - Schedule training rooms.
- One month before:
  - Confirm with staff which session they will be attending.
  - Review and revise documentation as necessary.
- On training day:
  - Employ interactive learning models so that staff benefit from hands-on experience.
  - Allow staff to engage in IM/SMS questions with "test" patrons:
    - One librarian can be the patron and the other can be the librarian.
    - Use these test examples to train staff on any unique features that the software may have.

Tip: If you are running a training session and your service is *live* you might see real questions come through the service from real patrons. A good practice to avoid missing the real questions is to preface all practice examples/questions as "test questions."

- Provide real-world examples of IM/SMS transactions in the training environment.
  - If you already have an IM/SMS service, use examples from this service or ask colleagues if you can use their transcripts (remove any identifying information).
- Provide staff with copies of documentation.
- Plan to hold refresher trainings annually.
- You may also want to provide training when needed (e.g., when new staff is hired).

For more information about how to implement technology training, see *Technology Training in Libraries* by Sarah Houghton-Jan (THE TECH SET #6).

CREATE AN INTERNSHIP PROGRAM

Up to this point staff have been defined as staff employed by your institution. Interns, often library school students, will now be included in this definition. An internship program is an excellent way to extend the hours of your IM/SMS service, and it allows master's candidates in library science (MLS or MLIS) to receive firsthand reference experience. However, if you do not have a university or college that supports an MLS program in your area, it is also possible to use undergraduates or graduate students from your institution or a local college or university in this role.

Alternative Options

Hiring MLS or MLIS students as interns is often an inexpensive and relatively easy option to add qualified staff to expand hours. However, for some institutions hiring interns will not be the preferred staffing model. Other options for extending hours include hiring adjunct librarians, creating a flex-scheduling option for librarians, or using a consortial staffing model as described earlier in this chapter. Depending on your institution, hiring adjuncts to cover evening and weekend hours may be acceptable, and there may already be funding in place for such staff. Flexible scheduling works well for those staff who are willing to work outside typical working hours. For example, staff could work 1 p.m. to 9 p.m. to cover an evening shift, or it might even be possible for them to complete their IM/SMS hours from home. It is important to remember that such a scheduling arrangement is dependent on agreement from the human resources department and managers of the IM/SMS staff.

The program must be well defined and branded before you can begin recruitment. Important questions that must be answered before you can begin include the following:

- Will the interns/students help extend service hours beyond what is already being provided, or will they fill in gaps during the existing service hours?
- Will interns/students receive compensation for their work, or can they receive course credit? (If you are hiring undergraduates or graduate students, you will most likely have to pay them in order to attract a good group of candidates.)
- What type of training will you provide? How often?
- How will you evaluate the interns/students over the course of time?
- Who will provide supervision of and training to the interns/students?

Your answers to these questions will help you form the basis for your intern or student program. If you do not currently have any program in place, talk to your human resources department so that you understand what the rules are for hiring students or for bringing on interns. It may also help to brand or name the program as well, especially if these positions will help you extend the service hours. For example, if your IM/SMS service is called "Ask a Librarian," then your intern program could be called the "Ask a Librarian Internship."

Recruit Interns

To find interns or students to staff the IM/SMS service, start by writing a detailed job description. The description should enumerate the responsibilities of the position, for example, "The intern will provide reference for a certain number of hours a week and participate in other projects as needed." In the case of MLS candidates who are using this for course credit, they will most likely be required to engage in some form of in-person reference or projects in addition to the IM/SMS reference responsibilities. When sending the description to local library schools, it should list the in-person projects and tasks in which the intern would be involved. For student employees you may be able to simply go through your human resources department and utilize the students hired in other departments or libraries at your institution.

Hire MLS Students as Interns

If you are able to create an internship program for MLS students to staff the IM/SMS service, then a more formal hiring process is required than when hiring student employees from your institution or from local colleges or universities. MLS students should be asked to submit a letter of reference with their résumé and cover letter. Depending on the number of applications that you receive, you may also want to conduct interviews with some of the students.

As part of hiring you must decide how many hours you will staff with interns. Interns are a good way to provide evening hours. Generally four hours per shift works well: an intern could staff 5 p.m. to 9 p.m. or 6 p.m. to 10 p.m. or some other variation of this. If you plan to provide evening hours every night, then you will need an intern for each evening, possibly two if you plan to offer long evening hours.

Another important consideration when hiring MLS students is their school's requirements for an internship. Many programs do not allow students to receive credit if an internship does not have an in-person component. In addition to the four-hour shifts for which interns will be responsible, MLS students will also have to work additional hours on campus with a librarian in order to receive credit. If you pay students, MLS or otherwise, to provide the IM service then this stipulation might not apply.

Sample Job Posting for Recruiting MLS Students

[Insert number] Positions Available

The Ask a Librarian Internship program provides current library school students with hands-on experience in several areas of academic librarianship. Interns will work collaboratively with mentoring library staff to assist users with the IM reference component of the Ask a Librarian service, answering IM questions from users.

Other instruction opportunities, such as writing instructional content and creating tutorials with the Camtasia software, may also be available as time permits. Additionally, interns could be assigned projects pertaining to collection development and library instruction, as well as participate in drop-in library workshops.

Interns will be appointed on a semester basis and may apply to renew the internship for up to two years. Preference is given to candidates who have completed at least 12 credits in an accredited library school program and have already taken their basic reference course.

POSITION AVAILABLE: [insert month and day], 20XX. Position is subject to renewal each semester.

WORK SCHEDULE: Work schedule is variable but requires at least one evening shift of four hours and six hours of on-campus working with a librarian for a total of 10 hours per week.

(Continued)

(Continued)

DUTIES: Interns will provide research assistance to individuals and groups via IM and refer users to appropriate staff from the libraries. Interns have the potential to teach drop-in library workshops to students, faculty, and staff and to create instructional content and tutorials for the web or inclusion in the course management system or even iTunesU.

REQUIRED: Must be current (or enrolled) graduate student in an accredited library school program during the [insert fall, spring, summer] 20XX semester.

PREFERRED: Demonstrated effective verbal and written communication skills. Course work in reference service and completion of at least 12 credits in an accredited library school program. Strong commitment to service. Ability to evaluate, understand, and respond to research questions by utilizing knowledge of library resources. Interest in academic librarianship. Ability to multitask and work effectively in a fast-paced, team environment. Experience using Camtasia software.

APPLICANT INSTRUCTIONS: Applicants for the Ask a Librarian Internship program should submit, via e-mail, a letter of interest, résumé, and at least one letter of recommendation by [XX] to: [persons responsible].

Train Interns

Once you select a candidate or a group of candidates, depending on how many hours you need covered, creating a training program will be critical. Typically this training can be done by your IM/SMS coordinator or by the intern's supervisor; in many cases, these may be the same person. However, undergraduates and graduate students will need the same training even though they do not have to go through the formal hiring process.

For students not affiliated with your institution, training can be more complicated, as they must become familiar with your institution's policies and procedures as well as your resources. It is especially important that these interns be properly trained if they will not be staffing the service after normal business hours. The same training protocol as outlined for staff can be followed here, but it will need to be slightly modified. For those interns who are not already part of your institution, a section should be added that emphasizes the internal policies and procedures of your institution. Good documentation on your institution's internal website will be critical for these staff. You may also want to consider getting your interns set up with access to your electronic resources remotely, as they will often be covering their IM/SMS hours from home or off campus and access to these resources is often critical in helping patrons with their questions.

Evaluate Your Interns

It will be important to provide consistent and frequent feedback to interns about their IM/SMS chats. If your system keeps transcripts, providing weekly comments on their interactions is a good way to do this. Problems can be identified quickly, and by providing frequent and detailed feedback you can prevent the interns from making the same mistakes over and over. If it is feasible, having regular meetings with them to provide feedback in person and to answer questions is a good way to keep up morale and find out if there are any other problems or concerns.

If your system does not keep track of transcripts, ask interns to copy and paste transcripts into a system such as Google Docs. Create an account that everyone can access so that interns can view each other's transcripts. Have intern supervisors provide comments directly on the transcripts within Google Docs. Even if your system keeps transcripts, this may be a good way to provide feedback to interns on their IM transactions; it can be a timely and simple way to monitor the progress. See below for a sample transcript from an intern and feedback provided by the supervisor.

Sample Transcript with Feedback

8:59 guest925967@chat.libraryh3lp.com: Are books listed as e-books not available in hardcopy?

8:59 intern: It's possible. Have you tried searching the CLIO record for the hard copy?

9:00 guest925967@libraryh3lp.com: Yes, I searched CLIO, and under Call Number it says e-book.

9:00 intern: In that case, the book is an e-book.

9:00 intern: You may be able to request it in hardcopy through ILL or Borrow Direct.

9:00 guest925967@libraryh3lp.com: Ok, thanks very much.

9:00 intern: No problem!

9:00 intern: Have a nice night!

9:00 guest925967@libraryh3lp.com: You too!

Great job! Keep up the good work!

▶ CREATE AND USE WIDGETS

Embedding chat widgets in a website provides the ability to more closely connect with your users. Chat widgets provide real-time feedback and discussions, a feature lacking in e-mail. However, there are also disadvantages: in some cases, widgets may slow down the time it takes the webpage to load or take away from the overall site experience rather than enhance it. Website design plays a key role here (see Chapter 6 for more on this topic). Not all websites may benefit from the inclusion of an IM widget, as some libraries may not have the time or resources to operate the technology to its fullest potential. However, it can't hurt to give a chat widget a try and make your own conclusions; the implementation of a chat widget should be analyzed on a case-by-case basis (adapted from Aidan Henry's "10 Chat Widgets to Consider," October 2, 2007, http://www.readwriteweb.com/archives/10_chat_widgets.php).

A list of chat widgets and descriptions of the most popular chat widget software is provided on the following website: http://www.readwriteweb.com/archives/10_chat_widgets.php. Use that discussion to help you decide between aggregator IM widgets. One of the most

Putting Webchat Widgets on Your Webpages

It's easy to get started! First create a widget using a code generator and then embed your new widget on your webpage. Widgets are usually customizable in terms of color and size. Different services and sites will have their own widget generators and provide different levels of widget customization. Figure 5.1 shows an example of the Meebo Me widget generator.

▶ Figure 5.1: Meebo Me Widget Generator

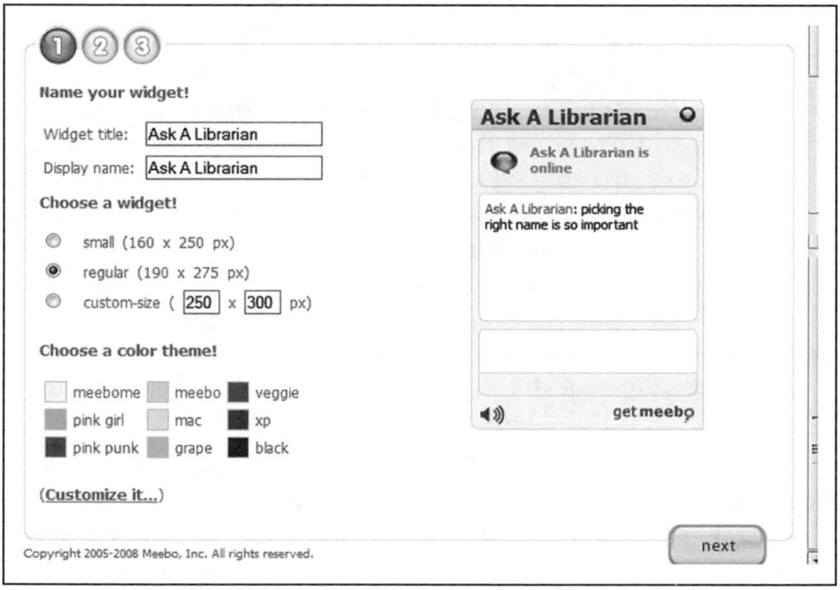

popular and commonly used in libraries today is the Meebo Me widget, which is free.

Basic Widget Design Tips

Generally you can customize the appearance of your IM widgets by altering their size, shape, color, and so forth. For example, the Meebo Me widget generator allows you to choose between two standard widget sizes or to design your own custom-sized widget based on pixels. It also lets you choose a color theme from six common color themes or customize your own. Think about customizing the widget to blend in with the color scheme on your library's website.

Many widget codes are "hackable," that is, manipulatable. Once you have seen how the HTML code works, you will be in good shape to manually edit it. You can tweak the HTML code to get the widget designed to your desired specifications. It is also often possible to customize what your users see when your chat service is offline. You might point to your e-mail reference service or to your 24/7 backup chat service (if you have one). In addition to customizing things such as color, size, and shape of your widget, some IM/SMS service providers (e.g., LibraryH3lp) offer different types of widgets for use on the web.

The basic types of widgets are embedded, pop-up, and follow-me (adapted from LibraryH3lp documentation at https://libraryh3lp.com/docs/which-widget.html):

- **Embedded widgets:** These sit inside a larger webpage and stay put within that page.
- **Pop-up widgets:** These widgets are chat boxes that open as a separate window when the patron clicks on an image (an arrow in the case of LibraryH3lp) that when you scroll over reads: "Opens chat in a new window." You can design your entry point to look however you would like. You can use text, or you can make customized online/offline buttons. See Figure 5.2 for an example of a pop-up widget.
- **Follow-me widgets:** This is a chat box that floats on top of the webpage, allowing the patron to continue the chat as he or she navigates to other pages.

For an overview of the pros and cons of each type, see Table 5.3.

Experiment with the different types of widgets and see what works best for your patrons and with your website. Also consider where on your website you are placing the widget. If you are in an environment where the patron will be clicking on a large number of links, such as

▶ Figure 5.2: Pop-Up Widget

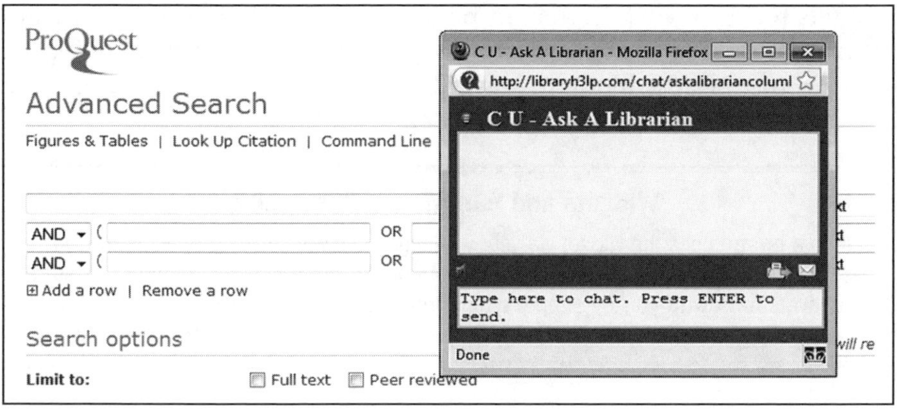

▶ Table 5.3: Pros and Cons of Using Widgets

Type	Pros	Cons
Embedded widgets	▶ High visibility; you will probably get more chats overall using embedded widgets. ▶ Links sent to a patron will be clickable by the patron if preceded by http://, and these links will open in a new tab or window. ▶ Embedded LibraryH3lp widgets automatically come with a "pop-out" feature to alleviate some of these issues, such as navigating away from the chat or the user having to scroll through too much text.	▶ It is very easy for a patron to accidentally navigate away from this type of widget by following links on the larger webpage. ▶ If the widget is too small it is hard for the patron to follow a conversation without a great deal of scrolling.
Pop-up widgets	▶ Patrons can easily flip back and forth between your chat and the other windows they are using for research. ▶ They always open in a new window and are resizable by the patron.	▶ Because the chat box is in a separate window, patrons will need to go back and forth from the chat box window to another window if they need to navigate the web during the chat. On the other hand, patrons who use IM are used to this.
Follow-me widgets	▶ The chat window stays on top, so patrons can always chat and continue with their research simultaneously. ▶ The "vanishing patron" problem is *greatly* reduced.	▶ Patrons receive a warning message before navigating away from the chat. ▶ Follow-me widgets are built on an iframe, and frame-buster websites can break the chats.

inside a catalog, you may not want an embedded widget. Embedded widgets tend to work better on pages for things like Ask a Librarian services or subject guides, where the page itself is for getting help; the patron may be less likely to start clicking other links on a page like this one and wander away from the chat session.

Widgets and Your Meebo Buddy List

When a patron uses the widget via your webpage, they will show up on your buddy list as new visitors or guests. You can IM visitors on your buddy list or visitors can start a conversation with you by simply typing into the text area on the widget, making this a really easy way to connect to your patrons and to provide reference services.

How Meebo Works

Meebo aggregates other IM services like AIM, Google Talk, and Yahoo! Messenger and lets patrons contact you using whatever IM provider they prefer to use or, if you have it set up, the Meebo Me widget. Librarians respond to patrons through the single log-on on Meebo's website: http://www.meebo.com/. Only one librarian can be logged on at a time. If another librarian logs on to Meebo, the previous librarian is bumped off, and any ongoing chat sessions will be lost. In order to prevent lost sessions, it is best to have a procedure in place in which the librarian who is getting ready to take over the next shift notifies the current librarian monitoring the IM service that he or she is ready to log on. Once the next librarian gets the message that he or she can log in, the librarian can proceed and continue monitoring the service. Librarians can contact each other by phone, by IMing the librarian through the widget, or through another IM account if he or she has one.

To create a Meebo Me widget:

1. Go to: http://www.meebome.com/.
2. Name your widget, and choose a display name.
3. Pick the size of your widget, or customize it. Click "Next."
4. Meebo generates the HTML code you need for your website; just copy and paste this code into your website's HTML code where you'd like the widget to appear.

How LibraryH3lp Works

LibraryH3lp allows multiple librarians to receive chats from its own chat widget or from Meebo Me widgets as well as receiving IMs directly from patrons on other IM networks such as AIM, Yahoo!, MSN, Google Talk, and ICQ. While it routes messages to multiple librarians monitoring the same queue, only the first librarian to respond "wins" the chat and becomes connected with the patron.

LibraryH3lp uses queues and gateways to manage and respond to incoming IM/SMS messages from patrons. Queues are public entry points: a queue is "online" to receive chats from patrons if at least one staff member is logged into the queue. LibraryH3lp gateways integrate public IM identities on networks such as AIM, Yahoo!, MSN, ICQ, Meebo Me, and Google Talk with a queue. Librarians can be assigned to queues based on their department or area of expertise. You can create as many individual user accounts and queues as you wish, and librarians can be assigned to more than one queue. Examples of queues are ask-a-librarian, science reference, humanities reference, subject-pages, course-pages, catalog, Refworks, and circulation.

To create a LibraryH3lp widget (you need a LibraryH3lp account to do this):

Note: If you chose LibraryH3lp as your software, continue with this section; if not, see the earlier Meebo section to create a Meebo Me account.

1. Log in to the LibraryH3lp admin site at https://libraryh3lp.com/admin/ and go to the "Design Widgets" tab in the LibraryH3lp admin interface to generate HTML to include widgets in your webpages.
2. Select the widget type you want to create: embedded, pop-up, or follow-me.
3. Use the default text ("JavaScript disabled or chat unavailable"), or write your own HTML to display when JavaScript is disabled.
4. Type in the HTML to display when no librarians are available, for example, "Chat is offline."
5. Next choose the Service Layer you are designing the widget for. You can design the widget for a particular queue, or individual librarians can have their own widget. Click "OK" and another tab will open where you can design and edit your LibraryH3lp widget. See Figure 5.3 for a screenshot of this. The new tab is where you can design and edit the features of your widget, such as:
   a. branding, title, text size, text color, typing identity;
   b. frame: boarder, width, height, background color;
   c. chat history: border, background, text color;
   d. message box: sent border, background color; and

► Figure 5.3: LibraryH3lp Widget Design Admin Screen

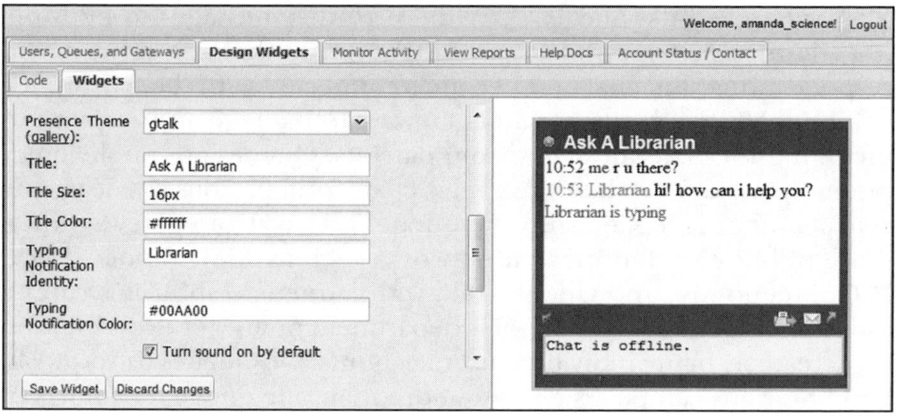

e. option to enter your CSS hack here or add another service layer (see later in the chapter for more on service layers).

6. The LibraryH3lp design tab displays your widget as you edit it, so you can see your work in progress and tweak it as you go. At this point it's also a good idea to save your widget skin so that you can come back to work on it again at a later point.
7. Once you are happy with the way your widget looks, click the "Generate Code" button, and LibraryH3lp will render the HTML code you can insert into the body of your website where you want your widget to reside.

You can incorporate your newly designed widget into an already existing webpage. Design the widget so that it fits with the design of your page. You can also design a new webpage around your widget. If you have an Ask a Librarian webpage, that's an ideal location for your library's widget. The LibraryH3lp widget code can even be "hacked" so that you can include branding specific to your library. It is possible to insert a banner into the top title portion of the widget. Directions for this can be found in the LibraryH3lp Help documentation: https://libraryh3lp.com/docs/customizing-widgets-by-hand.html.

Another option is to place the widget on a stand-alone page and then proliferate links to that page across your website. A widget permits anyone who has access to the webpage with the widget to communicate with the librarians and staff monitoring the IM/SMS service. If you want to restrict access to your widget, and therefore your IM/SMS service, you can put your widget behind a log-in screen where only authenticated users can access it. See Chapter 6 for more on the pros and cons of having your widget behind a log-in screen.

If you do not care about customizing the LibraryH3lp widget or what your patrons see when your widget is offline, you can embed your widget in an iframe and leave out all the rest of the code. Here is an example of iframe code:

```
<iframe src="http://libraryh3lp.com/chat/queue-name
@chat.libraryh3lp.com" frameborder="1" style="border:
2px inset black; width: 250px; height: 250px;">
</iframe>
```

This code is simple and works fine. LibraryH3lp defaults to hiding offline widgets and offering a true offline contact method, because they know from experience that patrons will try to chat with offline widgets, that they don't leave contact information, and that they will not wait for you to come back online.

You can design more than one type of widget for different parts of your website, and use them all! Examples include placing a follow-me widget in your library's catalog, creating an Ask a Librarian page with an embedded widget, and putting a pop-up widget in a subject-specific library guide.

How to Use Your Meebo Me Widget in LibraryH3lp

If your library has already produced and has been using a Meebo Me widget, you can keep using it with LibraryH3lp's routing system. To do this, just create a gateway for it, like you would any other IM account on the admin side of LibraryH3lp:

1. Go to the "Users, Queues, and Gateways" tab in the LibraryH3lp admin interface, and select which queue you want to assign your Meebo Me widget to.
2. Add a gateway by clicking on the "Add Gateway" button toward the bottom of the screen (see Figure 5.4).
3. Choose the IM protocol you want to create a gateway for (in this case, Meebo Me).
4. Enter the username and password you have previously created on the protocol's site. In this case, you will enter your Meebo Me username and password.
5. Click "OK," and you will have created a new gateway.
6. When your queue is online, your users will now be able to contact you through your Meebo Me widget.

► **Figure 5.4: LibraryH3lp Gateway Creator**

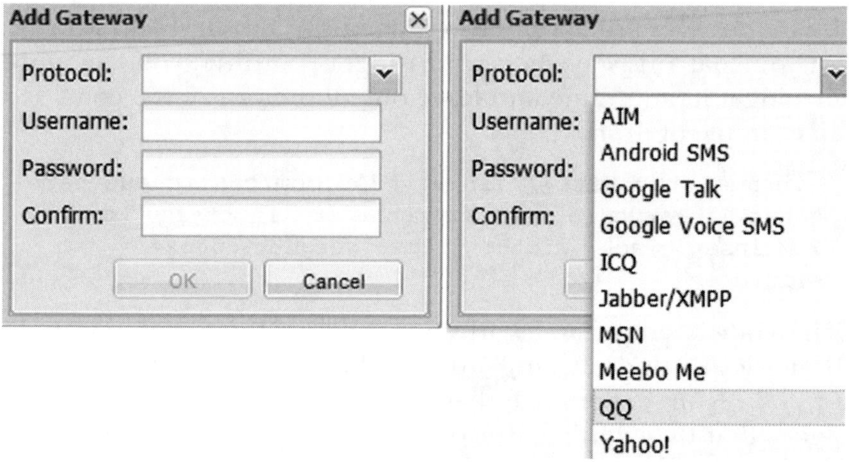

Tips:

- Do not delete gateways from a queue that is online within LibraryH3lp. Take the queue offline first.
- Once you have added a public IM identity using a gateway, your librarians should no longer sign into the public account with its AIM, Yahoo!, etc., username and password. The public identity will come online when operators assigned to the IM gateway's queue come online.

► USE GOOGLE VOICE FOR IMPLEMENTING AN SMS REFERENCE SERVICE

Google Voice gives you a single phone number that rings all your phones, saves your voicemail online, transcribes your voicemail to text, and allows you to send free text messages. Google SMS services work on every phone that supports SMS text messaging. Luckily, most cell phones these days are SMS/text capable. The basic features of Google Voice are free; however, for example, using Google Voice for making outgoing international calls has fees. Note: Patrons will accrue standard text messaging costs from their cell phone service provider.

Sending and Receiving SMS Messages Using Google Voice

It is possible to send and receive SMS messages through the Google Voice app for free. First create and activate a Google Voice account (see below) to respond to text messages from your patrons. To compose a text message, click the "Text" button at the top left on the Google Voice page (https://www.google.com/voice) above the Inbox. Enter the phone number, type your message, and click "Send." Replies work the same way by text message to a voicemail, call, or text: to text, click the "Text" link near the bottom of the message. Type your text message, and click "Send." You can also reply from your cell phone, but your mobile operator may charge you for it. Patrons can respond to your text from their phone, and you can respond from your Google Voice account and browser. Note: Only one librarian can be logged in to the Google Voice account at one time. Users can also use their own Google Voice accounts to contact you through the mobile Google Voice site at: http://www.google.com/mobile/voice/.

How to Set Up Your Free Google Voice Account

1. Create or log in to a Google account.
2. Go to: http://www.google.com/voice, and click the "Getting Started" link.
3. Choose your new Google Voice number. Use either words or an area code to find one you want to use. Once you have settled on a number you like, click "Continue." Choose a four-digit number to access your voicemail by phone. Once you accept Google Voice's Terms and Privacy Policy, click "Continue."
4. Enter a forwarding phone number that will ring when your Google Voice number is called. Click "Continue." Note: You may want to set up your Google Voice account to ring one or more of your public service desks in case someone calls it expecting to have a phone conversation. While you may advertise the number for your text message service only, once the number is in a contacts list, someone will very likely try to call it at some point.
5. As per Google's instructions, verify your phone. That's it! You now have a Google Voice account.

See http://www.google.com/support/voice/ for more help with Google Voice and Google Voice accounts.

Note: So far, Google has not shown signs of shutting down Voice applications or charging fees, but there is no guarantee this will always be the case.

Other SMS/Text Options

Now that you have created a Google Voice account you can configure other systems such as LibraryH3lp to work with Google Voice to route text messages through LibraryH3lp where librarians can respond to text messages from patrons just as they would any other message (see Figure 5.5). Using Google Voice with LibraryH3lp will allow you to enable SMS reference with no additional costs other than your LibraryH3lp subscription costs; however, your patrons will accrue standard text messaging fees from their service provider(s). Patrons will be able to send text message queries from their cell phone via your Google Voice phone number to your librarians via the LibraryH3lp interface. See the LibraryH3lp documentation for further instructions on how to do this.

Besides the Google Voice gateway, LibraryH3lp provides two other solutions that work with SMS. LibraryH3lp works with the Google Android phone (which your library would have to purchase or lease), which requires good network connectivity but would work internationally. For more on this, and to watch a video tutorial on how to set up your Android phone to work with LibraryH3lp, see https://libraryh3lp.com/docs/sms-gateway-android.html. The other SMS gateway that works well with LibraryH3lp uses Twilio. Twilio is a system that provides APIs and is intended to be used by developers for integration into applications like LibraryH3lp; and it is more robust than either of the other two LibraryH3lp gateways (Android phone or Google Voice). More about the use of Twilio with Library H3lp can be found here: http://libraryh3lp.blogspot.com/2010/06/super-cool-new-sms-gateway-twilio.html.

▶ **Figure 5.5: Text Message via LibraryH3lp**

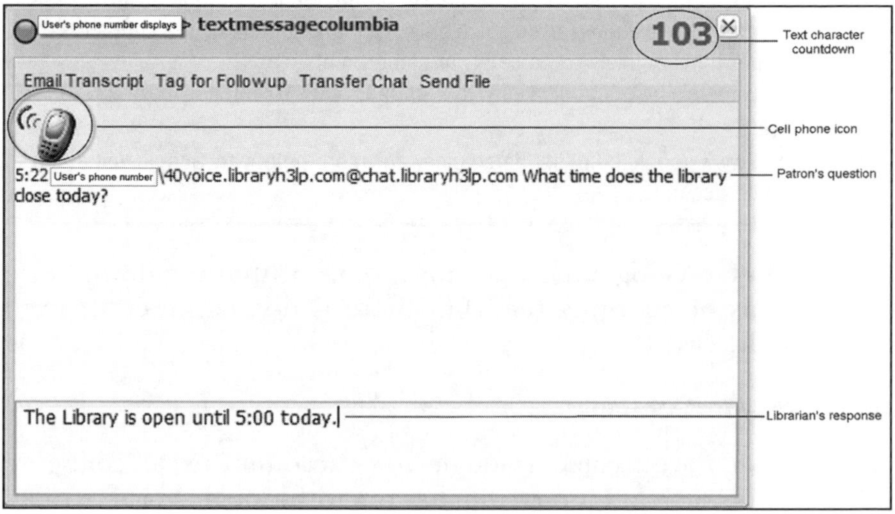

▶ UTILIZE AIM SMS HACK

Both LibraryH3lp and Meebo along with many other IM services work naturally with the AIM hack, which allows savvy AIM users to text AIM IM messages from their phones. For your users to use the AIM SMS hack, your library would not need to have a text messaging account set up—all you need is an AOL AIM account to monitor. With the AIM hack, patrons text their message to 246246. In the message, they specify your library's AIM buddy name. For example, if you monitor an AIM account in LibraryH3lp or Meebo, those text messages show up for all of your librarians just like any other traffic coming through the system. This hack also works from the librarian end of things as well. If your library wants to respond to text messages but does not want to have a cell phone, direct your users to use the AIM hack, and you can respond to their messages from within your IM software through AIM.

How Library Patrons Can Use the AIM Hack

1. Send a message to 246246 (AIMAIM).
2. Patrons enter the text in the following format: "send [library's AIM address] [question]" (without quotations or brackets); for example, "send anytownlibrary chat what are yr hours?"
3. From that point on, send and receive messages as you normally would. Your patron doesn't need to use the account prefix, and so forth, for each message during the session. Note that standard messages rates will apply.

To learn more about the AIM hack, see: http://aoltxt.com/im.

▶ RESOLVE SPAM ISSUES

With any public-facing IM/SMS service you run the risk of getting hit with some spam. Any instant messaging system (AIM, Yahoo!, MSN messenger) is subject to unsolicited messages, which could include commercial scam-ware, viruses, and messages with links for the purpose of fraud. If your IM reference service utilizes a public IM identity, your service could be at risk. There are some things you can do to combat spam. The first step is to learn to recognize when it is happening and what it looks like.

Recognize Spam

Spam messages come almost exclusively through public-facing usernames from public IM/SMS services like AIM or Yahoo!. Messages usually begin with a salutation like "heyyyyyyyyy" or "heyyyyyyy what's up????" Sometimes there are follow-up messages, frequently mentioning being bored (or, "sooooo borrrrrrred," as it were). There may also be messages about playing games or beating scores. In other cases they often send the same message repeatedly, such as, "you there?" Spam can also be unsolicited links to URLs or advertisements. These are not messages being sent by real people; they are spam.

Tips for Dealing with Spam

Make a note of the screen name that shows up in the chat window, and then close the window. Notify the person in charge of your IM/SMS service about the problem. Your IM system administrator should try to block the problem screen name from your IM service provider native interface. If most of your chat traffic comes in through widgets, you can still try blocking the problem IM accounts.

You may also consider reporting the spam to your IM service provider. Sometimes the IM service provider will be slow to block the spam bot or problem spam username. If you are using another IM service such as LibraryH3lp or QuestionPoint, you might also escalate the problem to the administrators of those systems to see if they can offer another solution. For instance, they might be able to block the problem username at a higher level. If these solutions do not fix your problem, you may want to consider temporarily disabling the particular IM service provider if it's really become a nuisance. Until the IM provider has had a chance to identify and block the spam bot, this may be your best option. While spam is both annoying and unprofessional, don't worry about it too much; it is, after all, just spam. Try to ignore these "junk" messages as best you can. Also, remember to remind your IM/SMS staff to try to have some patience until a viable solution is figured out. Keeping the staff in the loop about such things will go a long way to easing some potential frustrations.

▶ CREATE A MOBILE VERSION OF YOUR SERVICE

For more on mobile sites, see Jason Clark's *Building Mobile Library Applications* (THE TECH SET #12).

Mobile-Friendly Websites

If your administration allows it, you can design your library's website to be mobile friendly. Many web content management systems allow for the development of mobile-friendly sites; you can also adhere to some basic design principles to design your HTML code for a more mobile-friendly site. There are many new design features to keep in mind to make your site mobile friendly, and there are many websites that provide extensive advice and tips on how to accomplish this.

If your whole website is not mobile friendly, it is also possible to create a page or two that are mobile friendly. You can also create a pointer or a link, often a footnote, to the mobile-friendly version of your site, which would be readable for mobile users directing them to the mobile-friendly version of your site. Also, a variety of methods for automatic mobile device detection have recently become available, alleviating the problem of patrons having to find the mobile version themselves by instead automatically being redirected to the mobile-friendly site from their mobile device.

Mobile-Friendly Widgets

Many of the widgets mentioned previously, Meebo and LibraryH3lp, for example, are naturally mobile friendly. With a little work, widgets can be displayed on many mobile devices' browsers. The LibraryH3lp widget, written in JavaScript, has a mobile-specific version that is optimized for mobile browsers. The LibraryH3lp folks have tested it extensively on the iPhone and iPod Touch and on the native Android phone browser, and it should work well with them. Unfortunately, the BlackBerry system has proven to have more problems with displaying JavaScript. The LibraryH3lp mobile-specific widget is prefaced with m: http://m.libraryh3lp.com/chat/queue@chat.libraryh3lp.com.

As with the regular LibraryH3lp widgets, you can change the appearance of your mobile widget almost any way you would like. The easiest thing to do would be to use the Widget Designer on the LibraryH3lp admin site to create a custom skin using your preferred colors, title, and so forth. Then, after you generate the code, look for the raw widget URL (in the example above) and place the m: at the beginning of the URL. Of course, you can also always customize your widget HTML by hand if you prefer, as previously described in this chapter. More on the use of the LibraryH3lp mobile version can be found at http://libraryh3lp.blogspot.com/2010/01/improved-mobile-widget.html.

Meebo also offers a mobile-friendly version. Librarians monitoring your service or patrons using your service can access Meebo on their smartphone by simply pointing their browser to http://www.meebo

.com/. There, users will find an interface customized for their mobile browser and will be able to chat with staff monitoring your Meebo IM account through their browser on their mobile device. Meebo has recently developed apps for iPhone, iPod Touch, Android, and BlackBerry allowing for easier IM communication through mobile devices.

SMS

SMS is naturally mobile friendly. By enabling an SMS/text message reference service you will be allowing your patrons to send in questions via text from anywhere they have service on their text message–enabled cellular device. Patrons do not even need a smartphone; any text messaged–enabled cellular phone will do. Patrons can text in questions from anywhere they desire, both from inside or outside the library, as long as they have a signal.

Remember to inform your patrons of your SMS reference service in appropriate places. You can advertise your SMS service phone number in stacks areas where you might not have staff stationed or computer terminals available, so patrons can text in any questions they might have such as help finding a call number location. Advertise your SMS service in other appropriate, relevant places where patrons need help, and you will most likely drive more users to your service, as well as help them at their point of need. It is also important to remind your patrons that they will be charged the normal text message fees when they text in questions.

▶ CREATE AN IM LIBRARY KIOSK

It is possible to take your IM/SMS service one step further. Beyond access to your IM accounts, or queues online through your library's website or SMS reference, you can also establish a library kiosk if your library has the space and desire for one. You can set up a dedicated terminal (hard wired or WiFi enabled) that opens to your library's IM widget. You can also create a library guide or dedicated webpage that has your IM widget embedded in the help screen for patrons to use. Depending on where your kiosk is physically located, you might want to focus on different types of information on the help screen. You can also use widget features such as the service rollover feature in LibraryH3lp to create a unique queue in LibraryH3lp for your kiosk widget so that you know where your patrons are coming from.

At Bobst Library at New York University (NYU) they have created a virtual "help spot" on an upper floor that is no longer physically staffed by librarians (all reference services were merged to a centralized location on the first floor of the library). The dedicated terminal is well labeled with signage identifying it as a "Help Spot." When a patron approaches the terminal, the screen opens to an IM widget, which is embedded in a LibGuide. Patrons can then chat with the librarian or staff member monitoring the queue. This has been a relatively new approach for the NYU Library, since the reference services were all merged to the single service point on the first floor. Figure 5.6 shows the kiosk at the NYU Bobst Library.

▶ Figure 5.6: New York University Bobst Library Kiosk

Source: Photo by Amanda Bielskas.

▶ 6

MARKETING

- Develop a Marketing Plan
- Create a Brand
- Spread the Word in Your Organization
- Create a Main Reference Webpage
- Promote Your Library's Services

Marketing can take many forms, and you can create a very successful marketing campaign with little or no funding. Of course some financial backing is always appreciated. Do not underestimate the resources that are already at your disposal in your library. Using existing resources is one way to circumvent a low marketing budget.

DEVELOP A MARKETING PLAN

You should be able to develop a fairly simple marketing plan relatively easily. A basic plan should include the following points:

- **Target audience:** Who uses your service, or who do you want to use your service? (Some of this may come out of your assessment; see Chapter 3 for more information on this.)
- **Marketing objectives:** What do you want to achieve with this campaign?
- **Marketing strategies:** How do you plan to achieve your objectives?
- **Budget:** How much money will you need to achieve your goals? Your budget can include print advertising, contest prizes, give-aways, publications, printing, and so forth.

CREATE A BRAND

A brand is the personality that identifies a product, service, or company. Branding is widely adopted by businesses to promote their services and products, and libraries should also take advantage of such a tool. Branding your IM/SMS reference service can help users identify, recognize, and return to your service. Consider your library's brand or a brand for your IM/SMS reference service. Branding your library's identity and IM/SMS service will result in more recognition and feeling of identity. To succeed in branding you must understand the needs and wants of your customers and prospects. The background research on your users you have already done (see Chapter 3 for more information) is relevant here.

A successful brand:

- Delivers the message clearly
- Confirms your credibility
- Connects to your target audience
- Motivates
- Creates user loyalty

SPREAD THE WORD IN YOUR ORGANIZATION

Buy-in from your colleagues, staff, and administration is key. If you do not have support on multiple levels, chances are your brand launch will not get off to a great start. You need the support of your organization behind you, particularly to help spread the word about your IM and/or SMS reference service.

Take the time to remind your colleagues to market the service to their constituencies of students and faculty. Ask staff to mention your library's IM and/or text SMS reference service during their orientation and instruction sessions. They can also add information about the IM and/or SMS reference service to any handouts, pamphlets, and so forth that they distribute to their students at these sessions. Orientation sessions are also a great time to hand out SWAG (souvenirs, wearables, and gifts) if your marketing budget allows for it. SWAG are promotional items that often include the institution's brand. For our purposes they can include pens, mugs, and USB keys all labeled with the library's brand and IM/SMS brand. Handing out SWAG at opportune moments is a great way to advertise your services. In addition to orientation sessions, marketing your IM/SMS reference service during regular instruction sessions can also be successful. Targeting patrons when they need research assistance will most likely drive more users to your service.

CREATE A MAIN REFERENCE WEBPAGE

One of the most successful things you can do is to redesign or create the library's main reference webpage. If you are creating or rebranding the service, decide on a name for the service that will be applied consistently on all websites and promotional materials. "Ask a Librarian" is a commonly used name. It is also advantageous to create a logo for the service that could be proliferated around the website. A logo will ensure consistency for your brand and guarantee that it links to the correct page for the service. See Figure 6.1 for examples of some library Ask a Librarian logos and buttons.

A website redesign can greatly improve access to your IM/SMS reference service. Some quick ways to improve access to your IM/SMS reference service include the following:

- If possible, do not put your IM widget behind a log-in screen, where only authenticated users can access it.
- A unified page for all of your reference services is the easiest way for patrons to find the appropriate help they need. Aligning all of your services onto a single page gives users a clear picture of how to access help, whether it is in person, via consultation, phone, IM, or text messaging. Placing an IM widget on this page is also an easy way to drive usage to your IM service. (See Chapter 5 for instructions on how to make IM widgets for Meebo and LibraryH3lp.)

Figure 6.1: Ask a Librarian Logos

▶ Figure 6.2: Before Webpage Redesign

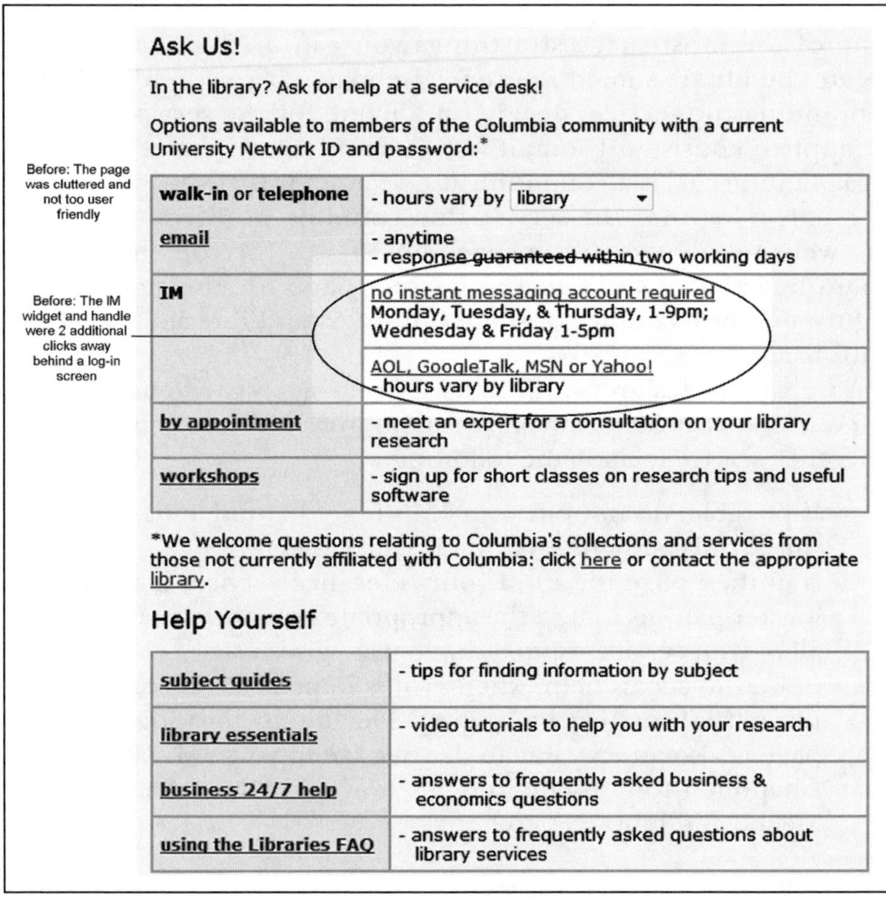

Figures 6.2 and 6.3 show screenshots before and after the redesign of the Ask a Librarian webpage on Columbia University Libraries' site.

Case Study

After redesigning the Ask a Librarian webpage and removing the log-in screen, staff at Columbia Libraries noticed a marked increase in their IM traffic. Traffic increased by well over 100 percent within the first month of the launch of the new page, and sometimes there is as much traffic in a single month as there was all year long with the old design. Staff attribute the greater increase in traffic to both the redesigned page and the ease of access to get to the IM widget.

▶ Figure 6.3: After Webpage Design

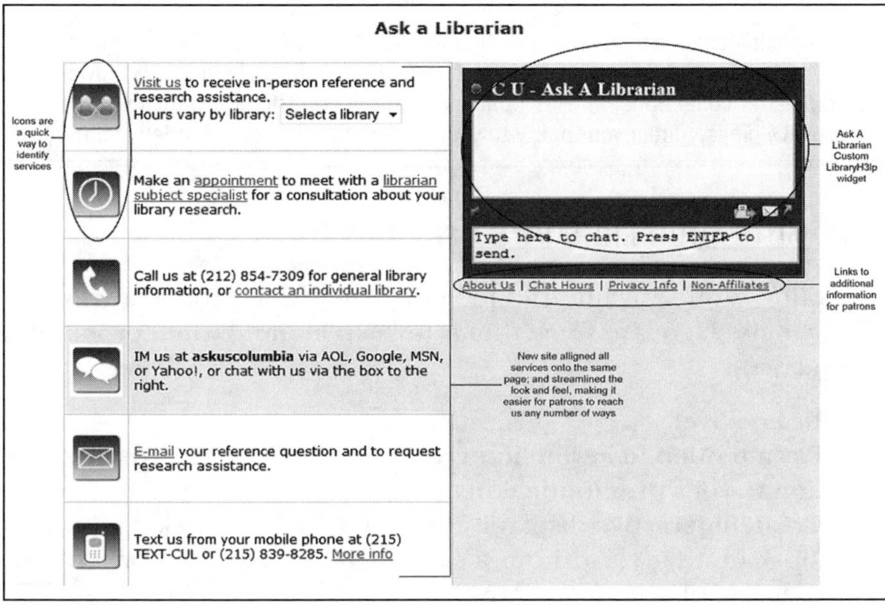

There are both good and bad reasons for and against having your widget behind a log-in screen (see Table 6.1). Most people against the idea of not having an authentication mechanism for users fear there will be hordes of people seeking free help. Both Yale and Columbia Libraries have an IM widget on their Ask a Librarian page that does not require users to authenticate before they can ask a librarian a question. Staff at both libraries verify that most users are affiliates of their respective schools and have legitimate questions. Both libraries have also added a disclaimer to their sites stipulating that help will be provided only to affiliates. See page 70 for an example of such a disclaimer.

▶ Table 6.1: Pros and Cons of Restricting an IM Service

Log-in	Pros	Cons
Required	▶ Limits access to authenticated users only. ▶ Only serious patrons make effort to ask questions.	▶ Adds layer of complication to website. ▶ Users may not make the extra effort necessary to ask a question.
Not required	▶ Anyone can access. ▶ Removes barriers to access.	▶ May open up the service to nonaffiliates.

Disclaimer

"Use of Ask a Librarian services is limited to John Smith University faculty, students, staff, and alumni; and to those with questions specific to the John Smith University Libraries, its collections, and its policies. If you're not affiliated with John Smith University, please direct your query to your local public, state, or academic library."

▶ PROMOTE YOUR LIBRARY'S SERVICES

Promotion is one way libraries let users know what products are available to them. Here are some things to keep in mind when promoting your services:

- **Be creative!**
- **Pay attention to aesthetics:** Use a variety of colors to draw attention to your promotional materials.
- **Create interactive displays:** You could create an Ask a Librarian kiosk in an area without a service desk (see Chapter 5 for an example of this at New York University's Bobst Library). All you need is a computer with Internet access; set the computer's home screen to open to the Ask a Librarian website or directly onto the chat widget.
- **Use events to promote your services:** Host a table at a student fair. During events held at the library, have promotional material available to be distributed or browsed.

Advertising Tips

- Make use of local school newspapers and student blogs.
- Advertise within the library by using signs and posting slides on LCD monitors.
- Integrate marketing for your IM/SMS reference service with anything your library is doing in terms of social networking, including Facebook, Twitter, and library blogs.
- Make use of print materials such as brochures, bookmarks, and handouts.
- Post flyers around the library.
- Put signage in appropriate areas where students might need help. If you are advertising a text message service, you may want to try posting signs in areas where there are no computers available and where students might want to text a question such as in the library stacks.
- Utilize Quick Response (QR) Codes. These shortcuts to websites are becoming common practice in advertising and are mobile-device friendly (see Chapter 9 for more on QR Codes).

Advertising can create new service opportunities. Don't be afraid to think outside the box!

You should also make use of the multiple avenues already available on the web to promote your services. For example:

- **Blogs:** Blogs are one way a library can market its services with little or no cost other than the time and talent it takes to create and implement the blog as well as the blog posts. Blogs can be used for brand awareness and to strengthen the brand.
- **Ambient reference:** Promotion of the product or service can be done by posting positive comments on other blogs as well as reacting to patron queries on topics dealing with the product or service. Does your school have an active blog where complaints about the library can sometimes be found? Taking the time to respond to these complaints and to offer possible solutions can go a long way toward building goodwill with patrons who often do not expect a response.
- **Social media:** You can make use of the same social media to attract new customers. In fact, this is a great place to spread the word on the product, thereby building the brand. Examples of such websites are the well-known Facebook, Flickr, and Twitter.

Limited Budget?

- Have local businesses donate prizes for contests.
- Ask a local business to donate printing services for newsletters.
- Hire interns for college credit. They can help you extend the service for additional hours, they get great experience, and patrons get help when they need it.
- Use staff with in-house expertise. Do you have a budding graphic designer on staff? Use her for help with designing graphics and logos.
- Join social media sites that are free (Flickr, Facebook, blogs, etc.) to expand your marketing reach beyond the traditional walls of the library.

▸7

BEST PRACTICES

- **Address Resistance to Change among Staff**
- **Establish Best Practices for Your Staff**
- **Establish Policies for Your Staff**
- **Establish Guidelines for IM Reference**
- **Use Canned Messages**
- **Employ Best Practices Consistently**

When establishing your IM/SMS reference service, it will be useful to the librarians and staff providing the service if best practices are codified. Best practices can relate to how questions are answered, when they get answered, and even how to deal with shift changes (if you are employing a scheduling process). Best practices should be provided on the website and/or through e-mail to all staff. This chapter presents some examples of best practices that can be amended for your institution and distributed to staff.

▸ADDRESS RESISTANCE TO CHANGE AMONG STAFF

Some public services staff may view reference in the IM/SMS environment as considerably different from reference that they are used to, that is, in person, via e-mail, and over the phone. Providing best practices that demonstrate how IM/SMS reference is similar to other reference services can help public service staff make the transition to this type of reference. These practices relate to how questions get answered and provide tips for librarians to handle different situations, such as nonaffiliates or multiple patrons:

- IM and SMS are new points of contact. Staff can apply policies already established for helping patrons in person, on the phone, or via e-mail. For example, if it is your library's/unit's/division's policy to not provide statistical information over the phone, for example, the current GDP for the United States, but to send a link to the appropriate government website, then follow the same policy when interacting with a patron on IM and/or SMS.

- For staff in private institutions who do not have a mandate to provide reference service to nonaffiliates, staff may want to refer nonaffiliates to their local public library (see later discussion on using canned messages, which would be appropriate for this situation; see also Chapter 6 for information about putting your service behind a log-in screen so that only current affiliates can log in).
- Stress to staff that they can take their time to think and type—often patrons are multitasking and do not expect an instantaneous response. If staff need time to work on a question, they can let the person know that they are still working on it or can offer to get back to him or her (see later discussion of canned messages for sample wording).
- Staff do not need to give out their names. Most software does not reveal internal IM/SMS handles, so patrons will only see that they are chatting with a librarian. Of course, if staff are comfortable giving out their name, they should. If your software has the ability to transfer questions to another librarian and you have decided to utilize this feature, it is a good practice to provide the name of the librarian to whom you are transferring the patron.
- Often a concern of staff is that they will be expected to adopt the IM or text speak that patrons utilize. Stress to staff that they are not expected to use these conventions, as it is important to convey a professional image. But it is fine to add a smiley face now and then! ;)
- Sometimes staff may have more patrons than they can handle (a sign of a successful service!). Staff should feel comfortable informing patrons that they are with another patron and will get back to them shortly (again, a canned message can be used here).

► ESTABLISH BEST PRACTICES FOR YOUR STAFF

In many cases best practices will depend on the software you choose. For example, some software may allow you to have multiple librarians logged in and to transfer questions to each other. Policies will need to be established for how this can be implemented and how librarians should handle the technology options. Here are some examples:

- With some software more than one person can be logged in at a time. If you decide to have multiple staff logged in at one time there are two possible ways to handle this:
  - If you are scheduling staff for discrete hours, only the person who is scheduled for that hour is responsible for answering questions; any other staff logged in can be available on a referral basis.
  - If multiple staff are logged in, no matter what scheduling method you use, whoever grabs the question first is responsible for answering it.
- If you decide that only the person who is scheduled for that hour should answer incoming questions, then the other staff logged in should be reminded not to pick up questions. Having multiple librarians and interns "fighting" for questions could lead to confusion.
- If your software allows you to transfer an IM question to another librarian, it is important that the librarian who is transferring the question do the following:
  - Contact the librarian to whom they want to transfer the question to make sure that he or she is available to receive the IM/SMS question.
  - Notify the patron that you are transferring him or her to another librarian.
- If scheduling staff for discrete hours and if it has been decided that only one staff member will answer questions during a shift, staff should get into the practice of handing off shifts even if the software allows multiple librarians to be logged in. Staff should IM the person scheduled before them to let him or her know that they are now available to answer questions.

ESTABLISH POLICIES FOR YOUR STAFF

Staff will need to work together to provide the best service possible. There may be technological limitations depending on the software you choose. If you are scheduling staff for discrete shifts, encourage them to be willing to pick up shifts when there are openings or when colleagues cannot cover their scheduled shifts. If you want to have a service where librarians transfer questions to one another, urge staff to be online as much as possible so that they are available for referrals.

Additionally policies should be implemented to ensure that the service runs smoothly and that all staff maintain a professional attitude. These include the following:

1. Staff log in for their discrete hours or shifts. When the next person takes over, that person should inform the previous person that he or she is ready to take over.
2. Staff should inform the person before them if they are going to be late.
3. Staff are responsible for finding coverage for their shifts when they are on vacation or have other commitments.
4. Staff should log out at the end of the day so that the service does not appear to be available when it really is not.

▶ ESTABLISH GUIDELINES FOR IM REFERENCE

Staff who are providing IM/SMS reference most likely provide reference service in a myriad of other ways and are well skilled at this aspect of their job. As stated earlier, many policies and standards can be applied or transferred to the virtual environment; staff may need to be reminded of this, however. The following best practices (pp. 76–78) are applied in Chapter 8 as guidelines for evaluating transcripts.

Sample Best Practices

The librarian responded to the "call" quickly.

Depending on your organization, you may want to include a measurement, for example, "quickly" means 60 seconds or less.

If the librarian could not help the person right away, he or she asked the user to hold on.

See discussion of canned messages for an example.

The librarian engaged in a reference interview.

This is the basis of the reference interaction and should be conducted over IM or text message whenever possible. IM/SMS staff shouldn't be afraid to ask questions and probe for additional information as they would when interacting with patrons in other formats.

(Continued)

(Continued)

The librarian was friendly and professional, avoided jargon, and did not use value judgments about the nature of the question.

This is an important reminder for librarians to behave as they would if they were interacting with the user through another medium, such as on the phone, in person, or through e-mail.

The librarian signaled an initial understanding of the user's needs.

This is particularly important in the virtual environment, as body queues such as nodding and eye contact or verbal queues such as "hmmm" are not available as they are in person or on the phone. The librarian must take care to inform the user that she understands the user's needs.

The librarian rephrased the user's question and asked for confirmation that the question was understood correctly.

Again, in the virtual environment this is particularly important, because the librarian does not have physical or verbal queues to aid in understanding. Rephrasing the question is a good way to make sure the user knows that you understand the question.

The librarian and user maintained a two-way conversation.

When a patron asks a question that requires research, the librarian should inform the patron that she needs to look up information rather than simply start researching without letting the patron know. This would be an example of a two-way conversation. Additionally, it is important that the librarians maintain their end of the conversation and do not ignore the patrons' inquiries.

The librarian sent messages to the user before sending webpages.

Because the librarian cannot use verbal or physical queues when interacting with the patron it is important that the librarian let the patron know when information is coming via the Internet, whether it be in URL or through the ability to push pages.

The librarian sent information in small pieces.

Users can digest information only in small chunks. If you are sending a great deal of information to the patron, then it is best to parcel it out. For example, if you need to send multiple URLs, perhaps send one at a time with a sentence or two of explanation.

The librarian offered some instructional guidance, including search strategies where appropriate.

As librarians we are charged with teaching users how to search. In the virtual environment this can be more challenging, but it can still be done. The librarian should try to include search strategies and provide information about how to search in the resource whenever possible.

(Continued)

(Continued)

The user indicated satisfaction before ending the "call."

The librarian should not just end the call but try to wait for a closing statement from the patron. This may not always be possible, as sometimes users will leave the chat and not signal that they are done.

The librarian referred the user to another librarian or library as appropriate.

Referrals are an important part of the reference transaction. If possible (and if your software allows this) transfer the patron to the appropriate person. Otherwise provide as much information as you can and let the librarian know to expect that the user will be contacting him.

Source: Adapted from Horde, L., and C. Tucker-Raymond. 2006. "Benchmarking Library Performance in Chat Reference." *Reference Librarian* 95/96: 5.

▶ USE CANNED MESSAGES

One way to ensure consistency among responses is to require (or encourage) the use of canned messages when possible. This can be particularly helpful when a patron initially messages a librarian or when a librarian needs to tell a patron to hold on. As a best practice it is important for librarians to learn how to use and get in the habit of employing them.

As part of using canned messages, you should indicate to staff which messages must be used during each interaction and which can be used at their discretion. For example, you may decide that the welcome message must always be used and that the hold message must be used when there is occasion for it. But the close message does not have to be used; you may want to encourage staff to provide a close message of their own choosing. The following are some pros and cons of using canned messages:

Pros:

- Staff can copy and paste responses for regularly received questions.
- Patrons always receive the same response to commonly asked questions.
- For new staff answering IM and/or SMS reference questions, having canned responses is often a welcome option.
- Canned messages are a good resource for finding answers to commonly asked questions.

Cons:

- Sometimes it can take longer to find the canned message than it would take to type out the response.

- Adding a canned message can interrupt the flow of a conversation.
- The nonpersonal nature of a canned message can make the conversation seem stilted and overly formal.
- Depending on the nature of your text messaging service, it may be difficult to send canned messages via text messages.

Sample Canned Messages

Nonaffiliates: Thank you for your message. [Insert library name] services are limited to [insert institution name] faculty, students, staff, and alumni and to those with questions specific to the [insert library name], its collections and policies. If you're not affiliated with [insert institution name], direct your chat question to your local public, state, or academic library.

Welcome/reading question: Hi! Thank you for contacting [name of your service]. I am reading your question and will be with you shortly.

Goodbye: Have a good day/evening! Please contact us again if you have additional questions.

Transfer: I am transferring your question to the [subject area] librarian. Please hold for a moment. [*Note: Remember to contact your colleague and make sure that she or he is available to accept the chat.*]

Switching shifts: My shift is over and another librarian is taking over. [If you can transfer: I will transfer our session to him now.]

Anything else: Is there anything else I can help you with today?

With another patron: I am assisting another patron at the moment. Please hold on and I will be with you shortly.

Hold: Please hold on a moment while I locate some resources to answer your question.

Follow up with more information: Please send me your e-mail address and I will send you more information.

Just a moment: Just a moment, please.

More information: Could you give me a few more details about what you are looking for and what you hope to do with the information?

More specific: Could you be more specific about what you are looking for?

Where looked so far: What resources have you consulted so far?

▶ EMPLOY BEST PRACTICES CONSISTENTLY

The transcripts in Figures 7.1 and 7.2 provide examples of how librarians can utilize canned messages and other reference best practices. Not every transaction will include all best practices, as there may not be an opportunity to conduct a reference interview or provide search guidance in all cases. Librarians will still have to use their skills and

► Figure 7.1: Chat Transcript: Example 1

7:59 guest11111@chat.libraryh3lp.com: what are the library's hours?

7:59 librarian: Thanks for contacting IM a Librarian. I am reading your question and will respond shortly. ← Librarian responds quickly and starts with opening canned message.

8:00 librarian: for which library do you need the hours?

8:02 guest11111@chat.libraryh3lp.com: the science library

8:03 librarian: great! The hours are 8am to 11pm daily. For more information see the library's website: www.library.science.edu ← Librarian provides information and URL.

8:06 guest11111@chat.libraryh3lp.com: OK, thanks very much! ☺ ← User indicated satisfaction before ending the chat.

8:06 librarian: no problem! Have a nice night! ← Librarian closes chat with own message.

8:07 guest11111@chat.libraryh3lp.com: you too!

► Figure 7.2: Chat Transcript: Example 2

11:29 guest12346@chat.libraryh3lp.com: where can I find market research reports?

11:31 librarian: Hi! What sort of market research reports do you need? ← Librarian responds quickly and tries to clarify the question.

11:35 guest12346: I am trying to find market research on coffee in different markets around the world. Do you have a database that has this?

11:37 librarian: thanks for clarifying. The database Global Market Information Database has such reports: [URL]. In the search box at the top right do a keyword search for coffee and USA, for example, to find reports on coffee in the US. ← Librarian provides a search strategy and URL for the database.

11:38 guest12346: OK - thanks very much ← User indicated satisfaction before ending the chat.

11:40 librarian: you are welcome! If you need more help please contact the business librarians at business@library.edu ← Librarian provides a way for patron follow-up and closes chat with own message.

11:44 guest12346: will do! Have a good night!

knowledge to decide what is best during each transaction so that they provide the best service. Still, some reference best practices should always be employed, particularly the following:

- Respond to the "call" quickly.
- If you can't help the person right away, ask the patron to hold on.
- Be friendly and professional, avoid jargon, and do not use value judgments about the nature of the question.
- Maintain a two-way conversation.
- If you have a lot of information to share, send it in small pieces.

In both examples the librarians responded to the patron right away, maintained a two-way conversation, and were friendly, professional, and did not reveal any value judgments about the questions. One initiated conversation by using a canned message while the other simply started with "Hi!" and tried to clarify the question. Both closed with their own closing message. The librarian in Example 2 took it one step further by providing a way for the patron to follow up for more information. Also in Example 2, the librarian was able to provide a search strategy (though a very simple one) to get the user started.

▶ 8

METRICS

- Measure Your Success
- Assess Your Service
- Determine Outcomes

Once your IM and/or SMS reference system is launched you will want to determine measures of success for the new service. Your definition of success will depend on your organization and what your goals were for the service. As discussed in Chapter 3, these goals will have grown out of the assessment you conducted before implementation. Establishing goals for your service is important, as they will help you evaluate its success and determine areas for improvement. Equally important is deciding what methods and metrics you will use to assess each goal. For example, if one of your goals is to increase awareness of the service, you will need to decide how you will evaluate that, an increase in usage perhaps, and then how much of an increase would indicate success.

▶ MEASURE YOUR SUCCESS

In Chapter 3 we developed three goals for the service:

1. Provide an IM and/or SMS reference service that is both easy for patrons to use and gives high-quality answers.
2. The service receives at least [insert number] questions in the first year with an increase by [X] percent in the second year. If you already have a service, the goal is slightly different: increase usage by [X] percent in the first year and [X] percent in the second year. [The metrics you use will depend on your organization.]
3. The hours meet the needs of patrons, that is, the service is available when patrons need to use it the most.

Using these goals, you can assess the success of your program. The first step will be to create metrics for each goal. See Table 8.1 for ways to measure these goals.

ASSESS YOUR SERVICE

Evaluating the usefulness of the service will be critical to its success. Just as we assessed the needs of users and librarians before introducing the service, you will need to measure the usefulness of IM and/or SMS reference at your organization so that it stays relevant to users and to the staff who provide the service. To begin, take the following steps:

1. Decide what you want to assess: all of your goals or just one or two.
2. Decide how you want to assess each goal, that is, a survey, focus groups, or other method (see Table 8.1).
3. If you have not already done so, create metrics for success for each goal so that you have a benchmark for your results.

Table 8.1: Assessing the Success of Your IM/SMS Reference Service

Goals	Metric	Method
1a. Provide an IM/SMS reference service that is easy to use.	X percent of users state that the process is seamless.	**Survey of users** Possible questions to assess this: ► How easy was it to find the IM/SMS service? (Rate ease of use on a scale of 1 to 5.) ► Did you have any problems using the chat widget or interacting with the librarian? (Rate on a scale of 1 to 5 or could be a free-text question.)
1b. Provide an IM/SMS reference service that gives high-quality answers.	For over X percent of questions librarians provide correct and friendly responses. (Note: This is more subjective, but you should use a high percentage, over 75 percent, for measurement in order to provide good service.)	**Survey of users** Possible question to assess this: ► Were you satisfied with the answer you received? (Rate on a scale of 1 to 5.) ► Was the librarian/staff member friendly? (Rate on a scale of 1 to 5.) **Review transcripts**

(Continued)

▶ **Table 8.1: Assessing the Success of Your IM/SMS Reference Service** *(Continued)*

Goals	Metric	Method
2a. The IM/SMS service receives at least [X number] questions in the first year, with an increase by [X] percent in the second year.	In first year: number of questions received. In second year: percentage increase.	**Analysis of usage data** Do these meet the goals you established before launching service?
2b. If you already have a service, the goal is slightly different: increase usage by [X] percent in the first year and [X] percent in the second year.	In first year: percentage increase in usage. In second year: percentage increase in usage.	**Analysis of usage data** Do these meet the goals you established before launching service?
3. The hours meet the needs of patrons, that is, the service is available when patrons need to use it the most.	Patrons are using the service heavily during evening and weekend hours.	**Analysis of usage data** Determine when patrons are using the service. **Survey of users** Possible questions to assess this: ▶ How satisfied are you with the hours? (Rate on a scale of 1 to 5.) ▶ Are there additional hours that you would like to see offered? (Free-text question.)

Assessment can include many pieces, including surveys or focus groups of users and librarians, review of transcripts, and evaluation of usage data. See Chapter 3 for sample focus group questions with staff; see page 86 for questions that can be used in a focus group with users. Assessment can be a lengthy process but it is worth the investment, as it will make for a better service all around.

Once you have done the assessment you will need to analyze the results and decide how to use the data. Never do assessment just for the sake of doing it! Make sure that you are collecting useful information that you can use to improve or change your service if that is what is needed.

Review Librarian Transcripts

Quality control is an important aspect of the evaluation of your service. If you have interns who are staffing the service you should have a

Questions Used During Patron Focus Groups

Warm up

1. How many of you have talked to or worked with a librarian since you've been at [insert name of your institution]? **Raise hands.**
2. Have you ever been disappointed by a service at the library? Tell me about it.
3. If you have asked a reference/research question at the library, tell me about that experience.
   a. What was your question?
   b. Were you satisfied with the response you got?
4. How do you typically use IM?
   a. Most of the time when you are on IM, who are you IMing?
   b. Do you ever IM professors, TAs, classmates, or people you work with? How often?
5. Which IM service do you use most? What do you like about it? What don't you like about it?

Questions

1. At what point in the semester do you usually use the library?
   a. What do you usually do at the library?
   b. Who do you usually go to for help with finding resources for your course work, like papers or projects? Why do you go to this person?
   c. If not a librarian, why?
2. How many of you were aware before this focus group that you could IM the library? **Raise hands.**
3. If you were aware that the library offered an IM service, is there a reason you haven't tried the service?
4. If you were to IM the library, who would you expect to be answering your question?
5. What time of day do you think an IM service is most useful? When are you most likely to IM a librarian?
6. Would you add the library to your buddy list? (Or would you prefer to go to the library's website to IM a librarian?)
7. Now that you know that the library has an IM service, do you think you'll use it? Why or why not? For what?
8. In general, what do you like about using IM? What do you dislike?
9. When would you use IM over other methods of communication in general?
10. Describe your ideal library IM service.
11. Can you describe what librarians do?

system in place to regularly review transcripts of their interactions and give them feedback (see Chapter 5, pp. 44–48, for more information). In terms of other staff it may be prudent to review their transcripts on a regular basis as well. While you do not want librarians and staff to feel that you are looking over their shoulder, establishing a procedure of regular reviews is useful to maintaining a service that is of the highest quality. Because you will most likely not be reviewing transcripts on a daily basis, spot-checking transcripts on a monthly, bimonthly, or even quarterly basis should be sufficient depending on the needs of your organization. (If your software does not keep transcripts, see Chapter 5, pp. 44–48, for a recommendation on how to solve this problem.)

Conduct Spot Checks of Staff Transcripts

Begin by developing a checklist of what you are looking for when reading transcripts to make sure that the samples you read are meeting the baseline level of quality (see below for a suggested list). Determine what would be a sufficient sample to read; for example, if you receive 300 IM and/or text message questions a month, perhaps you want to read 10 percent of the transcripts. Make sure that you randomly select which transcripts you read so that you remove any bias for or against certain staff.

Guidelines for Evaluating Transcripts

1. The librarian responded to the "call" quickly (depending on your organization you may want to include a measurement–for example, "quickly" means 30 seconds or less).
2. If the librarian could not help the person right away, he or she asked the user to hold on.
3. The librarian engaged in a reference interview.
4. The librarian was friendly and professional, avoided jargon, and did not use value judgments about the nature of the question.
5. The librarian signaled an initial understanding of the user's needs.
6. The librarian rephrased the user's question and asked for confirmation that the question was understood correctly.
7. The librarian and user maintained a two-way conversation.
8. The librarian sent messages to the user before sending webpages.
9. The librarian sent information in small pieces.
10. The librarian offered some instructional guidance, including search strategies where appropriate.
11. The user indicated satisfaction before ending the "call."
12. The librarian referred the user to another librarian or library as appropriate.

Source: Adapted from Horde, L., and C. Tucker-Raymond. 2006. "Benchmarking Library Performance in Chat Reference." *Reference Librarian* 95/96: 5.

Competencies for Librarians

As part of the evaluation of transcripts it is important to develop several key competencies for librarians who are answering questions via IM and/or text message. These can be discussed during training (see Chapter 5) and reinforced during transcript evaluation. Competencies should be a natural extension from the guidelines you use for evaluating transcripts and vice versa. The American Library Association's Reference and User Services Association (RUSA) suggests professional competencies for librarians and has published them in "Guidelines for Behavioral Performance of Reference and Information Services Professionals" (available at http://www.ala.org/Template.cfm?Section=Home&template=/ContentManagement/ContentDisplay.cfm&ContentID=26937). These can be useful to creating a list of skills that librarians should have for interacting with users in the virtual environment. Kwon and Gregory, in "The Effects of Librarians' Behavioral Performance on User Satisfaction in Chat Reference Services" (see Articles in "Recommended Reading"), indicate that librarians who use these guidelines when working with patrons via IM tend to have more satisfied clients. In particular, Kwon and Gregory note that staff who used the following behaviors were the most successful:

- Communicated receptively and listened carefully
- Searched with or for the patron
- Asked the patron if the question was completely answered
- Asked the patron to come back if he or she needed additional help

Luo, in "Chat Reference Competencies: Identification from a Literature Review and Librarian Interviews" (see Articles in "Recommended Reading"), categorizes important competencies into discrete areas, including reference interview skills, familiarity with electronic resources, customer service mentality, the ability to work under pressure, online communication skills, and basic computer skills.

DETERMINE OUTCOMES

Depending on the culture of your organization, reading transcripts of peers may be a politically charged undertaking. Having management support will be critical for the success of this initiative. Other critical components include the following:

- Clearly outline objectives and reasons for reviewing transcripts of peers, and communicate them at the outset. Reasons can include the importance of providing a high-quality service that users find useful, identifying if staff need additional support or

retraining, and helping librarians improve their skills through regular feedback. A good user experience should be your main objective, as you want to provide the best service you can.
- Tell colleagues what the outcomes will be for those who are not meeting the objectives so that they know what to expect. See the next list for possible outcomes.

See Chapter 4 for a discussion of communicating with staff.

Once you have done a few spot checks, or even possibly the first time, you may identify some staff who are not providing the level of service that you would like. Depending on the severity of the problems, there are several ways to handle them:

- You can retrain the staff member based on specific examples in the transcripts.
- Provide a mentor to the staff member who will read the transcripts daily or weekly (as appropriate) and provide regular feedback.
- Remove the staff member from the IM/SMS reference service.

Assessment is an important part of any library service. It helps you ensure that your users are receiving the best service and that staff have all the tools they need at their disposal while performing the service. Although informing colleagues their service is less than stellar is not easy, it will contribute to a better service now and into the future.

9

DEVELOPING TRENDS

- Investigate Online VoIP, Video, or Cobrowsing Reference Software Options
- Explore Mobile Applications

What's on the horizon for virtual reference services? What sort of new technologies will impact your services in the near and not-so-distant future? While it's almost impossible to predict the future, there have been some recent advances in technology that could impact areas in the realm of virtual reference services. Although we never know what the next technological advance will bring, here we try to sort out the exciting, up-and-coming technologies that might impact services in the virtual reference arena.

INVESTIGATE ONLINE VOIP, VIDEO, OR COBROWSING REFERENCE SOFTWARE OPTIONS

Many new software options on the market today use Voice over Internet Protocol (VoIP) and streaming and live video technology that allows you to keep in touch with your patrons in new ways. Several of these options are free; other software is available at a cost. Many of these allow for the integration with other technology, such as e-mail and IM. These software systems have not been designed with libraries specifically in mind but were developed for other purposes. That does not mean that libraries should not take advantage of these technologies; you might just have to be more inventive when you implement them in order to find a solution that will work best for your library. Here are a few examples:

- **Skype:** Skype is a software application that allows users to make voice and video calls and chats over the Internet. Calls to other users within the Skype service are free; calls made outside the

Skype network can be made for a fee. Additional features include instant messaging, file transfer, and video conferencing. Cost: mostly free; some calls cost extra.

- **Adobe Connect:** This is a software product used to create information and general presentations, online training materials, web conferencing, learning modules, and user desktop sharing. All meeting rooms are organized into "pods," with each pod performing a specific role (i.e., chat, whiteboard, note, etc.). Cost: fees apply.
- **LotusLive:** LotusLive offers e-mail, networking, and collaboration tools and online meetings so that people can work with others inside and outside of the organization. You can meet online, share files, chat, manage projects, network with potential clients, and send and receive e-mail anywhere, anytime. Cost: fees apply.
- **Windows Live Messenger:** This provides free voice, video, and sharing features. Users can share photos and videos while they chat. The video chat is in high definition, and users can send a video message to a friend. Cost: free.

EXPLORE MOBILE APPLICATIONS

Mobile Websites

More and more users have smartphones that allow them to access the Internet. Your library may want to consider launching a mobile-friendly website or create a subset of pages that are mobile friendly. A segment of those pages could be dedicated to your Ask a Librarian service. Many of the IM widgets mentioned in previous chapters also have mobile-friendly versions or are already mobile device ready. It would be possible to provide links to the mobile-friendly pages or widgets.

QR Codes

QR codes are basically fancy bar codes that are capable of holding more complex data and information than just numbers. Many smartphones that are equipped with a camera also have the ability to download a QR code reader, enabling the user to quickly scan a QR code and access the information that is transmitted by the code. QR codes can transmit different types of information, such as website links, phone numbers, or text numbers. Libraries can use QR codes as a quick and easy way for patrons to access the library's already existing mobile services.

Tablet Computers

iPads or tablet computers are relatively new to the market: the iPad was first released in April 2010. This evolving technology could have many possible applications for use with or in libraries. Along with the application (app) market, new uses for these devices are being invented on a regular basis. The iPad could be useful for virtual reference, especially for use with roaming reference services, where librarians walk around the library and offer to help patrons. As long as you have WiFi or 3G/4G access, your iPad could support IM reference and is both portable and mobile friendly.

Twitter Reference

Twitter is an online social networking/microblogging service that was launched July 2006. Twitter enables its users to send and read text-based posts of up to 140 characters, also known as "tweets." Some libraries adopted the use of Twitter early on as another communication tool with patrons. Since then, many libraries and librarians have begun to utilize Twitter for marketing, reference, and customer service purposes, among others. Libraries can use Twitter to locate people who are tweeting certain keywords, for example your library's name. Some libraries have also begun to use Twitter to tweet reference questions; for examples of this, search Twitter for #refdesk. Twitter accounts are free, so you might want to create an account and start experimenting with and adding it to your library's virtual reference portfolio. After all, 200 million-plus users can't be wrong.

Foursquare Strategies

Foursquare is a location-based social networking website based on software for mobile devices that was created in 2009. This service is available to users with GPS-enabled mobile devices, such as smartphones. Libraries might be able to take advantage of this technology to help bridge the gap between the real world and the virtual one. Libraries could use "check-in" locations around the library to advertise their services to other users based on the user's network of friends or followers. Libraries might also offer some type of reward system or prizes for patrons who check in a certain number of times or for the person who checks in the most often. For example, Harvard University has partnered with Foursquare to develop a game encouraging students to explore the university, rewarding them with special badges and points as they discover new places. Such a scheme could be used to encourage users to visit the library. (For more about Foursquare, see Melissa Rethlefsen's 2010 article "Checking In: Location Services for Libraries" at http://www.libraryjournal.com/article/CA6725234.html.)

Apps

Mobile applications, or apps, are software that is developed for small handheld devices such as a mobile phone or an iPad. These apps can be downloaded by users from various mobile software distribution platforms such as an app store or from a website. It is possible for libraries to develop and offer a variety of their own apps to their users. Many libraries are already offering apps for some services, including searching the library's catalog, finding library hours, managing library accounts, accessing special collections, and asking questions of librarians.

Google Maps

Libraries might want to develop some initiatives using the Google Map API or Google Earth. Integrating these tools into your virtual reference services might be tricky, but if you develop a useful tool, patrons will most likely adopt it and use it. Possible applications include maps of branch library locations and stack guides. The Google Map API also allows you to embed other types of data into your Google maps, such as book locations and locations of special collections. The street view application could be used to create a virtual tour of your library with links to additional information.

For more information on some of the topics mentioned in this chapter, see the other books in this series, including *Location-Aware Services and QR Codes for Libraries* by Joe Murphy (THE TECH SET #13) and *Building Mobile Library Applications* by Jason Clark (THE TECH SET #12).

So, whether your next steps beyond implementing an IM and SMS reference service include implementing a VoIP reference service such as Skype, using new technologies like an iPad with your reference service, inventing a "killer app" for your library, or some combination of these for something newer and even more exciting, stay abreast of technological advances. This will perhaps inspire you toward new areas in which to experiment and explore. Again, we never know what the next technological advance will bring, and you never know which future ideas and inventions will help you to improve services in the virtual reference arena.

RECOMMENDED READING

ARTICLES

Anderson, Jacqueline. 2010. "Understanding the Changing Needs of the US Online Consumer, 2010—An *Empowered* Report: How Online and Mobile Behaviors Are Changing." Forrester Research. December 13; updated May 18, 2011. http://www.forrester.com/.

This provides an excellent overview of consumer behaviors using online tools such as IM and e-mail.

Bishop, B.W. et al. 2006. "Virtual Reference Services: Consortium versus Stand-Alone." *College & Undergraduate Libraries* 13, no. 4: 117.

The writers studied the costs and benefits of providing a chat reference service at the University of South Florida.

Breidenbaugh, A. 2006. "Budget Planning and Performance Measures for Virtual Reference Services." *The Reference Librarian* 95/96: 113.

This article examines the budget implications for joining and continuing in a statewide virtual reference service.

Coffman, Steve, and Linda Arret. 2004. "To Chat or Not to Chat—Taking Another Look at Virtual Reference: Part I." *Searcher* 12, no. 7: 38–46.

The authors argue that marketing a chat service to increase service must also coincide with a decrease in costs in providing the service in order to make a case for continuing.

Cummings, Joel, Lara Cummings, and Linda Frederiksen. 2007. "User Preferences in Reference Services: Virtual Reference and Academic Libraries." *portal: Libraries and the Academy* 7, no. 1: 81, 86, 89–91, 94–96.

This study examines how a chat service is used in academic libraries and if it competes with other services such as e-mail or in-person reference.

Desai, Christina. 2006. "Instruction via Instant Messaging Reference: What's Happening?" *Electronic Library* 24, no. 2: 174–189.

The author argues that patrons need and want instruction even in the virtual reference arena, including IM.

Desai, Christina M., and Stephanie J. Graves. 2008. "Cyberspace or Face-to-Face: The Teachable Moment and Changing Reference Mediums." *Reference & User Services Quarterly* 47, no. 3 (Spring): 242.

This article examines the teaching role of librarians when providing reference by studying IM transcripts, e-mail transactions, and in-person interactions.

Duncan, V., and A. Gerrard. 2011. "All Together Now! Integrating Virtual Reference in the Academic Library." *Reference & User Services Quarterly* 50, no. 3: 280. http://www.rusq.org/wp-content/uploads/2011/04/Duncan-and-Gerrard.pdf.

In this study the authors describe how a comprehensive review of their offered reference services prompted systemwide changes to all of the library's reference services; the article concludes with a set of planning recommendations.

Eakin, L., and J. Pomerantz. 2009. "Virtual Reference, Real Money: Modeling Costs in Virtual Reference Services." *portal: Libraries and the Academy* 9, no. 1: 133.

This article provides a cost analysis for providing virtual reference services as part of a collaborative system.

Gilbert, L.M. et al. 2006. "Assessing Digital Reference and Online Instructional Services in an Integrated Public/University Library." *Reference Librarian* 95/96: 149.

The authors provide plans for current and future assessment of digital reference, including e-mail, IM reference, and online instruction.

Granfield, D., and M. Robertson. 2008. "Preference for Reference: New Options and Choices for Academic Library Users." *Reference & User Services Quarterly* 48, no. 1: 44.

The authors conclude that users prefer in-person reference, but virtual reference services such as chat provide an important service to populations such as graduate students.

Graves, Stephanie. 2006. "Instruction via Chat Reference: Does Co-browse Help?" *Reference Services Review* 34, no. 3: 340–357.

Graves argues that patrons want instruction even in the virtual reference medium and that more research needs to be done on using the cobrowse tool.

Henry, A. 2007. "10 Chat Widgets to Consider." *Read Write Web*, October 2, 2007. http://www.readwriteweb.com/archives/10_chat_widgets.php.

This is a good overview and comparison of embedded chat widgets.

Hyde, L., and C. Tucker-Raymond. 2006. "Benchmarking Library Performance in Chat Reference." *Reference Librarian* 95/96: 5.

This article describes how librarians in Oregon's collaborative statewide virtual reference project analyzes IM transcripts to evaluate the performance of librarians.

Kotter, J.P., and L. Schlesinger. 2008. "Choosing Strategies for Change." *Harvard Business Review* 87, no. 7: 130.

In this article the authors advocate for three different methods that can be employed to overcome resistance to change.

Kwon, Nahyun, and Vicki L. Gregory. 2007. "The Effects of Librarians' Behavioral Performance on User Satisfaction in Chat Reference Services." *Reference & User Services Quarterly* 47, no. 2: 137–148.

Even in the virtual reference arena the behavior of the librarian is important; this article discusses how librarians interact in this arena and if they are more effective if they follow the RUSA "Guidelines for Behavioral Performance of Reference and Information Service Providers" (which are available at http://www.ala.org/Template.cfm?Section=Home&template=/ContentManagement/ContentDisplay.cfm&ContentID=26937).

Luo, L. 2008. "Chat Reference Evaluation: A Framework of Perspectives and Measures." *Reference Services Review* 36, no. 1: 71.

This article attempts to provide a holistic framework for evaluating virtual reference services in multiple contexts.

Luo, Lili. 2007. "Chat Reference Competencies: Identification from a Literature Review and Librarian Interviews." *Reference Services Review* 35, no. 2: 195–209.

This paper identifies chat reference competencies and can be used as a basis for the design of training and education programs for chat reference librarians.

Meier, J.J. 2008. "Chat Widgets on the Library Website: Help at the Point of Need." *Computers in Libraries* 28, no. 6 (June): 10–13, 48.

This article describes the use of chat widgets on library websites and how they affect daily work and organizational culture.

Moyo, L.M. 2006. "Virtual Reference Services and Instruction: An Assessment." *Reference Librarian* 95/96: 213.

This paper assesses the incorporation of instruction in library virtual reference services and explores whether the rate and nature of instruction provided to patrons during sessions are different from those provided during face-to-face reference.

Naylor, S., B. Stoffel, and S.V. Laan. 2008. "Why Isn't Our Chat Reference Used More? Finding of Focus Group Discussions with Undergraduate Students." *Reference & User Services Quarterly* 47, no. 4: 342.

The authors discuss the findings of focus groups of undergraduate users to understand why chat reference is not used more.

Stoffel, Bruce, and Toni Tucker. 2004. "E-mail and Chat Reference: Assessing Patron Satisfaction." *Reference Services Review* 32, no. 2: 120–140.

Illinois State University librarians surveyed their e-mail and chat reference patrons via e-mail to determine how they feel about the services and how the services might be improved.

Taddeo, L.R. 2008. "U There? How to Reach a Virtual Audience through Affordable Marketing Strategies." *Internet Reference Services Quarterly* 13, no. 2/3: 227.

This article discusses the marketing strategies the University at Buffalo applies to chat reference, including advertising, use of icons, and user surveys for assessment.

Theiss-White, D., J. Dale, E.F. Melia, L. Bonella, and J. Coleman. 2009. "IM'ing Overload: LibraryH3lp to the Rescue." *Library Hi-Tech News* 26, no. 1/2: 12–17.

This paper provides an overview of the LibraryH3lp virtual reference platform, including what it does, how it works, and its benefits for reference managers.

Wan, Gang (Gary). 2009. "Key Issues Surrounding Virtual Chat Reference Model: A Case Study." *Reference Services Review* 37, no. 1: 73–82.

This paper investigates the role co-browsing plays in virtual reference services and provides a comprehensive method for transcript and usage data analysis.

Ward, D. 2003. "Using Virtual Reference Transcripts for Staff Training." *Reference Services Review* 31, no. 1: 46–56.

This article examines an alternative method of staff training in which graduate student workers at a university reference desk were asked to examine transcripts of actual virtual reference transactions using the standard of the Reference and User Services Association's "Behavioral Guidelines" (which are available at http://www.ala.org/Template.cfm?Section=Home&template=/ContentManagement/ContentDisplay.cfm&ContentID=26937) for evaluation.

Ward, David. 2004. "Measuring the Completeness of Reference Transactions in Online Chats: Results of an Unobtrusive Study." *Reference & User Services Quarterly* 44, no. 1: 46–56.

This article reports on the results of a study used to measure whether online chat reference can effectively answer short, subject-based research questions.

Warrick, D.D. 2009. "Developing Organization Change Champions." *OD Practitioner* 14, no. 1: 14–19.

The author provides strategies to develop "change champions" in an organization.

Wikipedia. 2011. "Brand Management." May 23. http://en.wikipedia.org/w/index.php?title=Brand_management&oldid=430366145.

This article discusses ideas of branding and brand management.

Wiley, Deborah L. 2008. "Virtual Reference Service: From Competencies to Assessment." *Online* 32, no. 4: 62–23.

The author reviews the book *Virtual Reference Service: From Competencies to Assessment,* edited by R. David Lankes, Scott Nicholson, Marie L. Radford, Joanne Silverstein, Lynn Westbrook, and Philip Nast, published by Neal-Schuman in 2007.

▶ BOOKS/DISSERTATIONS/THESES/REPORTS

Bishop, Bradley W. 2010. *Chat Reference and Location-Based Questions: A Multi-Method Evaluation of a Statewide Chat Reference Consortium.* Florida State University.

The author addresses a lack of knowledge about chat reference and location-based questions and the implications of this lack of knowledge on chat reference consortia.

DeAngelis, Jocelyn Aline. 2010. *Friction in Computer-Mediated Communication: An Unobtrusive Analysis of Face Threats between Librarians and Users in the Virtual Reference Context.* PhD, Rutgers University. http://hdl.rutgers.edu/1782.1/rucore10001600001.ETD.000053411.

This dissertation studies computer-mediated communication in which interpersonal communication content between library users and reference librarians who engaged in service encounters is evaluated.

Emanuel, J. 2004. *An Assessment of Chat Reference Usage and Trends at the Health Science Library of the University of North Carolina at Chapel Hill.* MSIS, University of North Carolina at Chapel Hill. http://etd.ils.unc.edu/dspace/bitstream/1901/70/1/jenniferemanuel.pdf.

The author examines the use of chat reference services in the Health Sciences Library at the University of North Carolina at Chapel Hill over a period of two and one-half years.

Harmeyer, Dave. 2007. *Online Virtual Chat Library Reference Service: A Quantitative and Qualitative Analysis.* Pepperdine University.

This study suggests a research-based conceptual accuracy model of best practices for appraising and improving virtual chat reference service.

Houghton-Jan, Sarah. 2010. *Technology Training in Libraries.* 1st ed. THE TECH SET #6. New York: Neal-Schuman.

This book outlines various types of technology training programs and helps you determine which one is right for your library.

King, M.M. 2009. *Is What You See What You Get? Exploring the Role of Virtual Reference Icons on Academic Library Websites.* MA, University of Alberta (Canada).

This study is meant to help library staff better understand the factors involved in icon design and better understand academic library design, selection, and use of web icons.

Luo, Lili. 2007. *Towards Sustaining Professional Development: Identification of Essential Competencies and Effective Training Techniques for Chat Reference Services.* University of North Carolina at Chapel Hill.

The author identifies the essential chat reference competencies and findings that can be used as the basis to design and implement training and education programs to enhance the professional preparation of chat reference librarians.

MARS Digital Reference Guidelines Ad Hoc Committee. 2004. *Guidelines for Implementing and Maintaining Virtual Reference Services.* Chicago: Reference and User Services Association (RUSA). http://www.ala.org/ala/mgrps/divs/rusa/resources/guidelines/virtual-reference-se.pdf.

These guidelines address implementing and maintaining virtual reference services.

Pew Internet & American Life Project. 2004. *How Americans Use Instant Messaging.* http://www.pewinternet.org/Reports/2004/How-Americans-Use-Instant-Messaging.aspx.

This Pew Internet research report discusses how Americans use instant messaging.

———. 2009. *Parent–Teen Cell Phone Survey.* http://www.pewinternet.org/Reports/2010/Teens-and-Mobile-Phones.aspx.

This Pew Internet research report discusses the prevalence of teen mobile phone use and texting.

———. 2011. *A Closer Look at Generations and Cell Phone Ownership.* http://www.pewinternet.org/Infographics/2011/Generations-and-cell-phones.aspx.

The generational use of cell phones was examined in this study.

Saunders, Laura. 2004. *Plan, Prepare, Promote: Marketing Your Library.* MLA Conference, http://www.masslib.net/conference/PlanPreparePromote.pdf. Accessed December 19, 2011.

This PowerPoint presentation explores marketing strategies for your library.

Wan, Gang (Gary), D. Clark, J. Fullerton, G. Macmillan, D. Reddy, J. Stephens, et al. 2009. *Key Issues Surrounding Virtual Chat Reference Model: A*

Case Study. Bingley, United Kingdom: Emerald Group Publishing. doi:10.1108/00907320910937299.

This study investigates the use of cobrowse in live chat, customers' question types, referral to subject experts, and patrons' usage patterns as experienced in the virtual reference chat reference services at Texas A&M University Libraries.

▶ WEBSITES

Adium. http://adium.im/.

Go to this website to download Adiuu, an instant messaging application for Mac OS X that can connect to AIM, MSN, Jabber, Yahoo!, and more.

AIM IM Hack (AIM TXT). http://mobile.aol.com/sms-txt/AIM-TXT/.

This useful website explains how to send a text that arrives as an instant message (also discussed in Chapter 5 of this book).

Altarama. http://www.altarama.com/.

This website is for the vendor of RefTracker, Deskstats, Refchatter, SMSReference, and VRLPlus.

Fire. http://fire.sourceforge.net/.

Go to this website for Fire, an instant messenger client for Mac OS X. Fire is based on freely available libraries for each service. Currently Fire handles AOL Instant Messenger, ICQ, MSN Messenger, Jabber, limited IRC, Yahoo!, and Apple Bonjour communications.

Google Voice. https://www.google.com/voice.

Go to this website to create a Google Voice account.

LibraryH3lp. http://libraryh3lp.com/.

This website is for LibraryH3lp, a live chat solution for individual libraries, library consortia, and nonprofits. It allows clients to unify chat across web, mobile, and IM services.

Library Success: A Best Practices Wiki. 2011. "Libraries Using IM Reference." Last modified September 16. http://www.libsuccess.org/index.php?title=Libraries_Using_IM_Reference.

This wiki was created to share ideas about how libraries and librarians are using technology.

Live!Zilla. http://www.livezilla.net/.

Go to this website to download the Live!Zillia platform. Live!Zilla provides IM and tracks usage of clients' websites.

Master New Media (blog). 2010. "Web-Based Instant Messengers: A Mini-Guide." Updated February 26. http://www.masternewmedia.org/news/2007/05/20/webbased_instant_messengers_a_miniguide.htm.

This website compares various IM platforms.

Meebo. http://www.meebo.com/.

Go to this website to learn more about the Meebo client.

Miranda. http://www.miranda-im.org/.

Go to this website to download the Miranda IM client, an open source client for MS Windows.

Pidgin. http://www.pidgin.im/.

Go to this website to download Pidgin, a free chat client that allows you to connect to AIM, MSN, Yahoo, and more chat networks all at once.

Plugoo. http://www.plugoo.com/.

Go to this website to download this free chat client.

Proteus. http://www.proteusx.org/.

Go to this website to download Proteus, a multiprotocol chat client that allows users to connect to several IM platforms at once.

QuestionPoint. http://www.questionpoint.org/.

Run by OCLC, QuestionPoint provides IM, e-mail management, and a chat widget. It also includes membership in the "24/7 Reference Cooperative."

Text a Librarian. http://www.textalibrarian.com/.

This is the website for Mosio's Text a Librarian service.

Trillian. http://www.trillian.im/.

Go to this website to download the Trillian IM client.

Twilio. http://www.twilio.com/.

Go to this website to learn more about Twilio, a web-service API with which developers can build communications applications for voice and SMS.

Upside Wireless. http://www.upsidewireless.com/.

Go to this website to learn more about text messaging solutions provided by Upside Wireless.

INDEX

N

P

Q

S

T

U

V

W

ABOUT THE AUTHORS

Amanda Bielskas is the Geology/Geosciences and Psychology Librarian at Columbia University and has been at Columbia since 2007. She earned her Master of Library Science at Long Island University, has a Bachelor of Arts degree in Environmental Geology from SUNY Plattsburgh, and has a master's degree in Anthropology from Hunter College. Currently, Amanda also serves as an Adjunct Librarian at New York University and The Borough of Manhattan Community College. Amanda has worked as the Senior Reference Librarian at the American Museum of Natural History and, in addition to her academic experience, has also worked in several public libraries on Long Island.

Kathleen M. Dreyer is Head of the Watson Library of Business and Economics. In addition to providing research assistance to students and faculty, Kathleen oversees the collections and staff of the library. Before coming to Columbia, Kathleen was a reference librarian at the Lippincott Library at the University of Pennsylvania and a collections assistant at the Jack Brause Library at New York University. Kathleen holds a Master of Science in Library and Information Science from Drexel University and Master of Arts in Quantitative Methods from Columbia University.